ENVIRONMENTAL SCIENCE, ENGINEERING AND TECHNOLOGY

MOUNTAIN ECOSYSTEMS

DYNAMICS, MANAGEMENT AND CONSERVATION

Environmental Science, Engineering and Technology

Additional books in this series can be found on Nova's website under the Series tab.

Additional E-books in this series can be found on Nova's website under the E-books tab.

Earth Sciences in the 21st Century

Additional books in this series can be found on Nova's website under the Series tab.

Additional E-books in this series can be found on Nova's website under the E-books tab.

ENVIRONMENTAL SCIENCE, ENGINEERING AND TECHNOLOGY

MOUNTAIN ECOSYSTEMS

DYNAMICS, MANAGEMENT AND CONSERVATION

KEVIN E. RICHARDS
EDITOR

nova publishers
New York

Library of Congress Cataloging-in-Publication Data

Mountain ecosystems : dynamics, management, and conservation / editor, Kevin E. Richards.
p. cm. -- (Environmental science, engineering and technology earth sciences in the 21st century)
Includes bibliographical references and index.
ISBN 978-1-61209-306-2 (hardcover : alk. paper)
1. Mountain ecology. 2. Ecosystem management. 3. Mountain biodiversity conservation. I. Richards, Kevin E.
QH541.5.M65M72166 2011
577.5'3--dc22

2011002568

Published by Nova Science Publishers, Inc. † New York

CONTENTS

PREFACE

Mountain ecosystems serve as early warning systems both for climate change and pollution by persistent organic pollutants, in particular because mountains are located near the sources and could serve as monitors of the levels of POPs in the environment. This new book presents topical research in the study of mountain ecosystems, including Euro-Mediterranean torrents; comparison of the impact of reduced rainfall on plant and soil processes in mountain and grasslands; atmospheric carbon dioxide transport over mountainous terrains and abandonment of previously managed meadows in mountain regions.

Chapter 1 - Intermittent streams, despite having stream flow only part of the year, play a dominant role in fluvial geomorphologic processes as well as the water balance equilibrium of mountainous and semi-mountainous areas, in all the environments of our planet. These streams also exhibit high morphological and hydrologic variability because they are present in different climate regions that range from arid and semi-arid to humid and very wet.

A distinguishable intermittent stream type is the "Euro-Mediterranean torrent." The term "Euro-Mediterranean" describes the geographic location they are present in. Specifically they are present in the: Iberian, Italian and Balkan peninsulas, Southern France and the islands of Sardinia, Corsica, Sicily, Crete and Cyprus.

The term "torrent" describes primarily intermittent but also ephemeral streams in mountainous or semi-mountainous settings that have frequent flash-flood events and carry large sediment loads. Their watersheds have steep slopes, high to medium stream bed gradients with remarkably high alluvial deposits. There are numerous torrents flowing through human infrastructures, urban areas, protected natural areas and archaeological sites throughout the Euro-Mediterranean region making their study, management and control essential.

The aim of this chapter is a detailed presentation of Euro-Mediterranean torrent case studies from its eastern (Greece) and western (Spain) regions. These will provide information to land and water managers on how to manage these torrents more efficiently and effectively.

A very important step to properly and effectively manage and control torrents is an analytical classification using morphometric, characteristics and Digital Elevation Models (DEM). This is the subject matter of the first case study that presents the classification done for torrents in Greece. with potential of being adopted to the entire Euro-Mediterranean area. This classification has the potential of being adopted for the entire Euro-Mediterranean region. The second case study demonstrates the use of Geographic Information Systems (GIS) in the identification of surface erosion potential at the watershed level in Greece. This will allow managers to concentrate their conservation efforts in areas that produce the greatest amount of sediments. The third case study is from Spain and deals with sediment transport. It presents the model USPED that has the capacity to incorporate both erosional and depositional processes in torrent watersheds providing a more accurate picture of sediment transport processes. Finally, the last case study also from Spain demonstrates the successful efforts made to mitigate the damaging effect of torrents though civil engineering structures and reforestation.

Chapter 2 - Whitebark pine (*Pinus albicaulis*) functions as both a keystone and foundation tree species in upper subalpine and treeline forest ecosystems throughout western North America. Populations of whitebark pine are declining nearly rangewide from a combination of threats, including the invasive fungal pathogen *Cronartium ribicola* (which causes white pine blister rust), widespread population upsurges of the native mountain pine beetle (*Dendroctonus ponderosae*), altered fire regimes and advancing forest succession, with further declines expected from climate warming. *Cronartium ribicola* is geographically the most widespread threat and could greatly reduce regional populations by itself; but, in some locations, such as the Northern Rocky Mountains, interactions of several threats are leading to especially rapid declines.

Interspecific interactions drive two key ecological processes within whitebark pine communities–seed predation and seed dispersal. Whitebark pine obligately depends on Clark's nutcracker (*Nucifraga columbiana*, Family Corvidae) for dispersal of its large, wingless seeds. Prior to seed dispersal,

North American pine squirrels (*Tamiasciurus hudsonicus* and *T. douglasii*, Family Sciuridae), efficient and voracious arboreal seed predators, typically cut down whitebark pine cones for storage in middens. These cones are essentially unavailable to nutcrackers, and hence their seeds rarely lead to regeneration. The authors present evidence that as whitebark pine declines within subalpine forests, the rate of predispersal seed predation accelerates and the probability of seed dispersal decreases, leading to reduced tree regeneration over time. The authors further show that the dynamics of these interspecific interactions depend on forest composition, and under certain conditions result in a positive feedback scenario that hastens the decline of whitebark pine populations. The current management goal to restore and maintain these valuable high-elevation whitebark pine communities centers primarily on the propagation of rust-resistant genotypes within populations, which is accomplished either through natural seed dispersal from resistant parent trees, or nursery production and outplanting of resistant seedlings. The authors discuss how their findings could be incorporated into a restoration strategy for whitebark pine.

Chapter 3 - In this chapter the role of mountain ecosystems in the dynamics and fate of Persistent Organic Pollutants (POPs) will be described. POPs are a group of chemicals released into the environment by anthropogenic processes, characterized by an elevated persistence, high bioaccumulation and toxicity. Recent research has shown an increasing trend of POP levels in mountain ecosystems all over the world, in various environmental media such as soils, ice, snow, vegetation and animals. Short term POP accumulation in snow and ice may occur in different mountain systems depending on atmospheric circulation, temperature patterns and precipitation rates. On a long term basis accumulation of POPs in mountain lake sediments may be used for reflecting temporal and spatial trends in deposition. Passive air sampling and bioindicators are now being used for demonstrating altitudinal trends, where cold condensation processes may cause higher levels of some POPs at higher altitudes. Geographical trends will be also discussed, considering research performed in the major mountain ranges around the world (the Alps, Pyrinees, Andes, Rocky Mountains etc.).

Chapter 4 - Socioeconomic changes in Central Europe have been accompanied by abandonment of previously managed meadows in mountain regions. Among others influences, alteration of N-cycling through such a land-use change may be reflected in N rhizodeposition including proteinaceous compounds. In this work the authors have determined no significant ($P>0.05$) effect of 15 year abandonment of previously long-term mown mountain meadows on total protein flow in the form of water-soluble root exudates in mineral soil of 3–8.5 cm depth. Throughout the vegetation season, 6–21 mg N per square meter was deposited every sampling occasion in exuded total proteins on both types of meadows.

Chapter 5 - The authors investigated trace gas transport by daytime mountain induced circulations based on aircraft data collected during the Airborne Carbon in the Mountains Experiment (ACME04) conducted in 2004. Using measurements of traditional meteorology and trace gases concentration associated with ecosystem, the authors found evidence of thermally induced vertical mountain circulations consisting of an upslope flow, a horizontal return flow, and a descending flow along the Front Range of Colorado. Turbulent air was generated by convective buoyancy along the mountain slopes and ridges, and by strong shear at the mountain ridge tops. The deep turbulent mixed layer was subsequently advected by the ambient large-scale and return flows across the plains east of the Front Range. The descending flow over the plains increased the atmospheric stability, which reduced the vertical mixing above the convective boundary layer over the plains. The mountain circulation and its interaction with the ambient flow and convective turbulent mixing effectively transported the air with its unique _E-mail address: jsun@ucar.edu (Corresponding author) 2 Jielun Sun and Stephan F.J. De Wekker characteristics of the CO2, water vapor, and CO concentrations in the lowest layer over the plains to above the convective boundary layer over the plains. This study suggest that mountain-induced circulations have significant impacts on estimates of the regional ecosystem-atmosphere carbon exchange.

In: Mountain Ecosystems
Editor: Kevin E. Richards, pp. 1-44
ISBN 978-1-61209-306-2

Chapter 1

EURO-MEDITERRANEAN TORRENTS: CASE STUDIES ON TOOLS THAT CAN IMPROVE THEIR MANAGEMENT

***Dimitris Emmanouloudis*[1]*, *José L. García Rodríguez*[2]†, *George N. Zaimes*[3]‡, *Martín C. Giménez Suárez*[4]# and *Evangelos Filippidis*[5]¶**

[1]Professor, Management of Mountainous Waters Labaratory,
Department of Forestry and the Management of Natural Resources,
University of Kavala Institute of Technology (UKIT), Kavala, Greece
[2]Professor, Hydraulics and Hydrology Laboratory,
Forest Engineering Department, ETSI Montes at the Polytechnic
University of Madrid, Ciudad Universitaria s/n, 28040, Madrid, Spain
[3]Lecturer, Management of Mountainous Waters Labaratory,
Department of Forestry and the Management of Natural Resources,
University of Kavala Institute of Technology (UKIT),
Drama Annex, 1 km Mikrohoriou, Drama, Greece

---

* E-mail: demmano@teikav.edu.gr.
† E-mail: josel.garcia@upm.es.
‡ E-mail: zaimesgeorge@gmail.com.
# E-mail: martincgs@ingenieros.com.
¶ E-mail: efilippidis@gmail.com.

[4]Forest Engineer, Hydraulics and Hydrology Laboratory, Forest Engineering Department, ETSI Montes at the Polytechnic University of Madrid, Ciudad Universitaria s/n, 28040, Madrid, Spain
[5]PhD, Department of Geoinformatics and Surveying, Technolgical Education Institute of Serres, Terma Magnisias, 62124, Serres, Greece

ABSTRACT

Intermittent streams, despite having stream flow only part of the year, play a dominant role in fluvial geomorphologic processes as well as the water balance equilibrium of mountainous and semi-mountainous areas, in all the environments of our planet. These streams also exhibit high morphological and hydrologic variability because they are present in different climate regions that range from arid and semi-arid to humid and very wet.

A distinguishable intermittent stream type is the "Euro-Mediterranean torrent." The term "Euro-Mediterranean" describes the geographic location they are present in. Specifically they are present in the: Iberian, Italian and Balkan peninsulas, Southern France and the islands of Sardinia, Corsica, Sicily, Crete and Cyprus.

The term "torrent" describes primarily intermittent but also ephemeral streams in mountainous or semi-mountainous settings that have frequent flash-flood events and carry large sediment loads. Their watersheds have steep slopes, high to medium stream bed gradients with remarkably high alluvial deposits. There are numerous torrents flowing through human infrastructures, urban areas, protected natural areas and archaeological sites throughout the Euro-Mediterranean region making their study, management and control essential.

The aim of this chapter is a detailed presentation of Euro-Mediterranean torrent case studies from its eastern (Greece) and western (Spain) regions. These will provide information to land and water managers on how to manage these torrents more efficiently and effectively.

A very important step to properly and effectively manage and control torrents is an analytical classification using morphometric, characteristics and Digital Elevation Models (DEM). This is the subject matter of the first case study that presents the classification done for torrents in Greece. with potential of being adopted to the entire Euro-Mediterranean area. This classification has the potential of being adopted for the entire Euro-Mediterranean region. The second case study demonstrates the use of Geographic Information Systems (GIS) in the identification of surface erosion potential at the watershed level in Greece. This will allow managers to concentrate their conservation efforts in areas that produce the greatest amount of sediments. The third case study is from Spain and deals with sediment transport. It presents the model USPED that has the capacity to incorporate both erosional and depositional processes in torrent watersheds providing a more accurate picture of sediment transport processes. Finally, the last case study also from Spain demonstrates the successful efforts made to mitigate the damaging effect of torrents though civil engineering structures and reforestation.

1.0. EURO-MEDITERRANEAN TORRENTS

1.1. The Euro-Mediterranean Region

The Euro-Mediterranean Region includes the southern shores and islands of Europe along the Mediterranean Sea. Specifically, parts of the Iberian, Italian and Balkan peninsulas and the large island of Corsica, Sardinia, Sicily, Malta, Crete and Cyprus (Figure 1) [1]. It represents the largest area worldwide with Mediterranean climate. The 15 countries in this region, starting from western Mediterranean shores and moving to the eastern are: Portugal, Spain, France, Italy, Malta, Slovenia, Croatia, Bosnia and Herzegovina, Montenegro, Albania, Greece, F.Y.R.O.M., Bulgaria, Turkey (European part) and Cyprus (Figure 1).

The Mediterranean climate of this region is characterized by warm to hot, dry summers and mild to cool, wet winters [2]. This region receives the majority of its yearly rainfall during the winter season, and may go for many months during the summer without having any significant precipitation events. Consequenthly precipitation is highest in December and January with its average approximately 100 mm. In contrast, July and August are the driest months with average precipitation as low as 10 mm. Summer precipitation events are typically thunderstorms. Temperatures have a comparatively small range between the winter low and summer high because most areas are near a large body of water (the Mediterranean Sea). The mean monthly temperature high is 30°C in August and the low below 5°C in January. Still there are areas with winters that are cold with annual frosts and occasional snowfall.

The climate of these areas is characterized as "temperate Mediterranean" and is typical of the central and northeastern Iberian Peninsula, southeastern France, away from the immediate coastline, northern Italy, and northern Greece.

Figure 1. The Euro-Mediterranean region.

In contrast there are other areas that experience rather high temperatures in the summer such as southern Greece and Cyprus. The climatic characteristics of an area depend on its distance from the open sea, elevation, and latitude.

The region has a complex topography [1]. While it is primarily along the Mediterranean seacoast with many flat areas it also includes many large and tall mountain ranges. These include the Sierra Nevada and Pyrennes in the Iberian Peninsula, the Apennines in the Italian Peninsula and the Dinaric Alps, Pindos and Phodope in the Balkan Peninsula.

The climate and topography of the region along with the historical anthropogenic pressure dictate its diverse vegetation. It consists of extensive woodlands communities dominated by both evergreen and deciduous tree species and diverse evergreen shrublands, especially oak (*Quercus* spp) [3]. In the relatively moist areas there are tall evergreen shrublands (called *maquis*) and woodlands often with pines (*Pinus* spp). Dryer areas are dominated by shorter shrubs that are often aromatic. These areas are called *garrigue*. Finally, in the lowland arid areas in the eastern Mediterranean Basin spiny shrublands often occur that are called *phrygana*. Overall the Mediterranean vegetation is fire-prone with frequent fires occurring especially during the hot dry summer season. Fires are is a major problem of the region [3].

Finally humans have lived in this region for thousands of years [3]. As a result the natural environment has been strongly impacted and these impacts have continued to accelerate over the last century. Many consider the Mediterranean region the most impacted in the world by humans [4].

1.2. What are the Euro-Mediterranean Torrents?

A torrent is a type of stream. The term originates for the Latin word *torrente* and is commonly used in Italy, Spain and Greece. In most cases, it is an intermittent stream although there are torrents that are also ephemeral streams. In *intermittent* streams, water flows for part of the year while in *ephemeral* streams water flows only during and/or immediately after precipitation events [5].

In other parts of the world other terms are used to describe intermittent and/or ephemeral streams [6, 7, 8]. Specifically, in Latin America the term *arroyo* is used, in Britain the term *winterbourne* while in the Arabic-speaking world's the term *wadi* is quite common. Wadi's also includes ephemeral streams. Finally, in the southwestern United States the term *wash* is often used for ephemeral streams. Each term describe streams developed in areas with specific climatic and topographic characteristics that also lead to specific hydrologic characteristics. So while these terms describe intermittent and ephemeral streams they are not synonyms to torrents because of the significant hydrologic differences.

It is also important to understand the differences between torrent and river flow [8, 9]. Torrent flow is irregular and more difficult to predict than river flow that is smooth and more regular (no sudden extreme water flow changes). Torrent flow is faster with shallower water depth but greater sediment transport capacity compared to river flow. The higher transport capacity is the result of the greater slope of torrent channels compared to river channels. Torrents channels have slopes more than 2% while river have slopes less than 1.5%. Finally rivers have greater water depth and a significantly larger number meanders.

Torrents during the majority of the year do not have "torrential flow" in the dramatic sense of the word. There are even periods in which the flow is reduced to a trickle or less. Still it will have periods with large volumes of water and material rapidly flowing (torrential flow) through their channels. This is what characterizes them. The characteristics and the behavior of each torrent differ from regions to region depending on the combination of climate, geology, topography and vegetation of its watershed [9].

Euro-Mediterranean torrents are found primarily in mountainous and semi-mountainous areas flowing through small valleys and ravines with many ending in the coastal areas. Watershed areas are typically less than 150 km^2 with abrupt relief and steep slopes. Their maximum discharges occur during the spring and autumn and cause extreme, sudden and short in duration floods. These floods can carry great amounts of material that are autochthonous (eroded from the adjacent surrounding areas) that are deposited eventually in the flatter areas. Their increase in water flow originates from precipitation events and surface runoff and occasionally because of snowmelt. In contrast, during the hot and dry summers, the torrent flows are minimal and can have significant periods of time with no flow. During periods of no rain, the water flowing in their channels originates from groundwater. The Euro-Mediterranean torrents are developed in material with high porosity that allows quick recharge of the groundwater during and after precipitation events. Finally, its water quality is affected by channel, watershed (e.g. soil, slope and groundcover) and precipitation characteristics and differs from season to season but also along the length of torrent.

Overall the most important characteristics of the Euro-Mediterranean torrents are:

- *Irregular yearly water flow.* The flow can vary from no flow during the summer to extreme floods after precipitation events or snowmelt. Although these flood events last for short periods of time they can cause significant damages.
- *High sediment transport capacity.* Significant loads of material that are produced in a great variety of erosional processes within their watersheds are transported through torrent channels. The material typically is eroded from the mountainous and semi-mountainous areas and deposited in the flatter coastal areas.

1.3. Benefits and negative impacts of torrents in the Euro-Mediterranean region.

Water scarcity is a major concern in the Mediterranean region that impacts hydrologic resources, biodiversity, water quality, river and stream ecosystem functioning and the entire population of the region [3]. The Mediterranean Basin also is one of the most vulnerable regions in the world to climate changes and has also been impacted by human water demand the most [2].

Torrents are essential for the Euro-Mediterranean region since they are the most common stream type. As a major part of the water cycle they recharge the groundwater aquifers [5, 9]. The riparian areas along the torrents, are essential wildlife habitat, serve as corridors for fish and wildlife migration, connect fragmented habitats thus conserving biodiversity [9, 10].

In this water-scarce region it is surprising that efforts, to utilize torrents water flow (a renewable resource) beneficially are minimal to non-exist. With proper management plans torrent water flow can be stored and provide another source of water that could be used for agriculture, municipalities and even for restoration, conservation and protection of the environment. This will be crucial for the region with the impending climate change impacts on its water resources [2].

One of the major reasons for the minimal efforts of exploiting torrents is their irregular flow and high sediment transport capacity. This irregular flow of torrents can lead to sudden large with short duration floods, flash floods [7, 8]. These flash floods have destroyed throughout the regions cities and towns, transport routes, agricultural crops and even caused the loss of human lives. There are also other indirect damages primarily because of the large amounts of material that torrents have the capacity to transport. The large amounts of material they carry degrade the water quality of rivers and fill up reservoirs decreasing alarmingly fast their water capacity. Finally, because many of the torrents flow directly to the sea they can pollute it but also fill in sea ports.

1.4. Objectives of the Book Chapter

Torrents water flow is a valuable natural renewable resource that has not been exploited but can also cause very serious damages. It is imperative that torrents need to be better managed to both, be better utilized and minimize their damages. In the following chapters, case studies are presented from the eastern (Greece) and western (Spain) regions. One of the case studies from Greece presents a successful classification system of torrents for Greece that could be adopted for the entire Euro-Mediterranean region. The second case study from Greece, provides an example on how using Geographic Information Systems can indicate the watershed areas with the greatest erosion potential remotely. As for the Spanish case studies, one provides a model that in addition to erosion also incorporates deposition that allows the better understanding and estimation of sediment transport in the torrent channel and its watershed. The other case study demonstrates various management techniques practices to control the damaging torrential flow. Overall, this chapter presents new tools (e.g. a classification system, GIS and a modeling technique), along with an actual demonstration of an operational torrent management system, that could help land and water managers to effectively and efficiently manage torrents.

2.0. Case Studies from Greece: Torrent Classification and Estimating Potential Watershed Erosion Using GIS

2.1. Introduction

The Greek peninsula has several large rivers that flow through it but there is an even greater number of streams with most them being torrents. Kotoulas [9] reported that Greece has more than 1000 torrents. The Greek term for torrent is *cheimarros* (spelled with Greek alphabet letters as "χείμαρρος"). Cheimarros originates from two Greek words, *cheimerios* that means during the winter and *rous* that means flow. In other words cheimarros, refers to the streams that flow during the winter time. In Greece most precipitation events occur in the winter and early spring the period that torrents have their major flood events.

The reason for the dominance and development of torrents in Greece is the Mediterranean climate and the topographic and geologic conditions. Despite the significantly large number of torrents more emphasis has been given to the larger rivers in Greece. However, with the idea of the holistic watershed management approach, to effectively manage large rivers, there is a need for proper management of the torrents feeding these large rivers. Therefore a better understanding of torrent functions along with the use of novel tools and methodologies for their management is very important within the framework of the holistic approach..

Despite the small size of the Greek peninsula, many different types of Mediterranean climate exist. These range from the dry Mediterranean areas that can be found in the southeastern parts of Greece and the Aegean islands, to the humid Mediterranean areas in the northern and western parts of Greece and to the alpine Mediterranean areas that can be found in the tall mountain ranges such as Pindus and Rhodope. This highly influences the vegetation and the flow and types of rivers, streams and torrents of the peninsula [9, 11].

The abrupt topographic relief of the peninsula is also very influential because it leads to very extreme and dangerous flows particularly in the mountainous regions where torrents are more prevalent. These flows can carry large amounts of sediment and deposit them in the flatter coastal regions of Greece corroborating the relationship between mountainous and coastal regions. Overall as you move from the mountainous regions to the coastal ones, the torrent stream flow and sediment characteristics change substantially.

Torrent flash flooding is quite frequently in Greece causing substantial economic damages and in some cases even the loss of human lives. Mitigating torrent floods should be a priority in Greece. One of the major obstacles is the great variability of torrents flowing through the Greek peninsula that means specialized and localized plans need to be developed in order to effectively manage each torrent. This is extremely time consuming and very costly. A solution to this problem is the classification of torrents that can lead to the development of management plans specific to each type of torrent. A classification scheme for the torrents of Greece is described in the first case study. The second case study deals with the use of new technology to estimate the potential of erosion in a torrent watershed. Erosion can degrade the watershed ecosystem while the transportation and deposition of large amounts of sediments is a major problem caused by torrents. Indentifying areas with high erosion potential is a prerequisite in order to effectively manage torrents. This allows targeting the most erosive areas and minimizing torrent flow sediment loads.

2.2. Case Study 1: Classification of Greek Torrents

2.2.1. Background

The great variety of climatic, topographic and geologic conditions of the Greek peninsula has led to a plethora of torrents with unique and individual characteristics. After many years of working and surveying torrents in Greece it became apparent that despite their individuality many had common characteristics that could lead to their classification. This current effort of classification builds on a previous effort by Kotoulas [12] who tried to classify torrents in Northern Greece.

Many other sciences (e.g. botany, zoology, geology) have made substantial efforts on classifying plants, animals, rock formations etc. These classifications have helped improve the scientific study in their field and the communication among scientists of different countries. In the field of hydrology there have been some efforts to classify rivers and streams although none have been universally accepted. The greatest problem in trying to develop such a classification is the great diversity of streams. One of the most detailed classifications has been developed by Rosgen [13] for the Western United States.

In this case study an effort was made to develop a national scale classification scheme of torrents in Greece. In order to accomplish this aim, 172 torrents were selected and their characteristics were recorded. Two criteria were used to select the torrents: a) a uniform geographical distribution of torrents in all the regions of Greece (Figure 2) and b) a normal distribution of the watershed areas of the torrents (Figure 3).

2.2.2. Methodology

The key to a successful classification is using the appropriate parameters that best characterize the variable of interest. Of course if you use too many parameters then it looses its practicality. In the past, classifications in natural resources typically used morphometric and/or physiographic parameters.

For this classification morphometric parameters were used. There is a plethora of morphometric parameters (watershed area, channel length, channel frequency etc.) that describe torrents behavior which refers to how active the torrents are e.g. their frequency of flow, flooding and causing damages, their erosion and sediment transport capacity. The decision on what variables to use was based on a previous study conducted by Emmanouloudis [14]. The Emmanouloudis [14] study found that the morphometric variables that best descried the behavior of torrents in Greece were the mean slope of the watershed (Jm_K) and the mean slope of the torrents main channel (Jm_Λ). Other researchers have also found that Jm_K and Jm_Λ are good indicators of the torrent's behavior [15]. These two parameters are good indicators of the topography and the geology of the torrents watershed [16]; both very influential factors in torrent development. These two slope parameters also incorporate indirectly the torrents watershed vegetation that is typically an expression of its geology and topography [17, 18]. Vegetation cover and type highly influence surface runoff and erosion that impact torrent flow and flooding.

Figure 2. The regions of Greece.

To simplify things even more, a quantitative indicator was developed that included both slopes and was symbolized with the Greek letter Φ. The mathematical relationship between Φ and the two slopes is:

$$\Phi = \sqrt{(Jm_K * J\mu_\Lambda)} \qquad \text{(equation 1)}$$

The Φ indicator should be able to describe torrent behavior because of its direct relationship with Jm_K and $J\mu_\Lambda$. High Jm_K and $J\mu_\Lambda$ values result in high

Φ values implying that the torrents are highly active while low JmK and JμΛ values result in low Φ values.

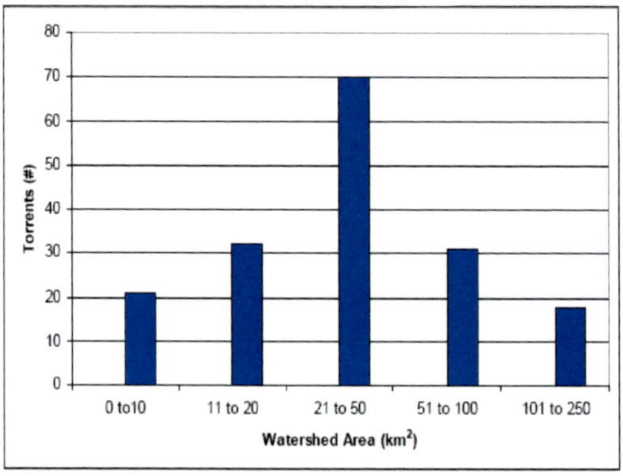

Figure 3. The number of torrents surveyed for the classification. The torrents were grouped in the following 5 categories based on their watershed area: a) very small 0-10 km^2, b) small 11-20 km^2, c) medium 21-50 km^2, d) large 51-100 km^2 and e) very large 101-250 km^2.

Through a statistical analysis it was found that the Φ indicator also has a good correlation (R = 0.90) with the mean slope of the deposited material of the torrent ($Jm_{\alpha\pi}$) [14]. This slope is a very important morphometic parameter in order to describe torrents. Aulitzky [19] states that the volume, shape and slopes of the torrents' deposited material help understand its long-term behavior and dynamics.

Thus the Φ indicator despite being a single number provides information directly or indirectly on the torrents general condition, its channel and its deposits, the topography and the vegetation of the watersheds, all essential morphometric characteristic that describe its behavior.

The final and most critical procedure was to validate the acuracy of the Φ indicator in describing torrents and more importantly torrent behavior. To accomplish the validation approximately 80 torrents were surveyed in the field while an additional 92 were investigated by utilizing topographic, geologic and vegetative maps.

2.3.3. Classification

The process to develop the categories of torrents involved a number of different steps. The first step was the establishment of the boundaries, the highest and lowest numerical values of Jm_K and Jm_Λ of the torrents. This was done by examining the values of the 172 torrents that were surveyed in the field or through the thematic maps. Based on these values, the higher and lower slope values for Greek torrents were indentified as:

min $Jm_K \approx 2\%$ max $Jm_K \approx 35\%$

min $Jm_\Lambda \approx 20\%$ max $Jm_\Lambda \approx 60\%$

These values can be perceived as representative due to the large number of the torrents studied (172) as well as due to the fact that the selected torrents are well-distributed both throughout the regions of Greece and from a watershed area standpoint (Figure 3).

The next step was to develop the boundaries of the torrent categories. To accomplish this the numeric interval by which Jm_K and Jm_Λ values would increase from one category to the next needed to be determined . After careful consideration it was decided that the interval would be 5%, except for the first Jm_K interval where it was only 3%. The interval of 5% was chosen because a smaller interval would increase the number of categories substantially. Once the interval was determined all paired combinations of Jm_K and Jm_Λ intervals (e.g. Jm_K: 5-10% is combined with Jm_Λ: 20-25%) were formed. From these combinations only the ones that were found in actual torrents were kept. These combinations were 45 and provided the initial torrent categories. Each initial category had a minimum and maximum Φ value.

Forty five categories was quite a large number of categories. In addition, when looking more carefully at the torrent categories there appeared to be substantial overlap of the Φ values between some of the initial 45 categories. This led the authors to reconsider and develop a more concise classification. In order to accomplish this, previous combinations with substantial overlap of the Φ indicator were grouped. This was done by maintaining the interval of 5% for Jm_K, except for the first interval that was 3% (2-5%) and the last interval that became 10% (25-35%). In contrast, the Jm_Λ interval was expanded to 10% when its values where greater than 25% and less than 55%. This Jm_K interval was maintained because it is more sensitive in predicting torrent behavior than Jm_Λ. It must also be noted that while the last Jm_Λ the interval was only 5% (55-60%) its respective Jm_K interval was quite wide (15-35%). When the watershed has very steep slopes (greater than 55%) they have the same or an even greater influence on torrent behavior than Jm_Λ.

This second grouping led to the final 20 torrent categories. While some of these categories have the same Jm_K or Jm_Λ, they always have different ranges of Φ values. Each category was ennumerated with a capital letter of the Latin alphabet beginning form A (Table 1). The great diversity of torrents in the O and N categories led to the development of two subcategories for each (Oi and Oii; Ni and Nii).

The final step to validate the classification was to examine the other physiographic and morphometric characteristics of the torrents in each category. A short description of some characteristics of each torrent category is provided in Table 2 (Figure 4). The characteristics described are:

- *Watershed area:* A range of torrent watershed areas are given for each category. Watershed area is important because it influences the total amount of surface runoff that can accumulate in the torrent channel.

Table 2. The most common geologic formations in Greece

Symbols	Name
f	Flysch
l	Limestone
d	Dolomite
c	Conglomerate
m	Marls
s	Slates
gn	Orthogneiss
sgn	Paragneiss
g	Granite
n	Other igneous
pg	Pleistocene deposits

- *Watershed relief:* The watershed relief is reported by the largest hypsometric difference of the watershed. This is estimated by feature by subtracting the highest elevation of the watershed, typically a peak near the headwaters and the lowest, typically the mouth of the torrent. *Geologic formation:* The dominant geologic formations of the watershed area are reported because it relates to the watershed and channel slope and the erosivity of the watershed. In some watersheds more than one geologic formation was present. The most common geologic formations in Greece are presented in Table 2.

- *Vegetation*: The type of vegetation(s) and in some cases the dominant species are described. In many cases different vegetation types were dominant particularly between the upper and lower parts of the watershed that had a large hypsometric difference
- *Depositional formation:* The shape and/or the slopes of the alluvial deposits were reported since they are highly influenced by the torrent's activity [14].
- *Region:* The geographic region(s) each category of torrents is found in Greece is also reported. When a region(s) is not specified, the torrents of the particular category present throughout Greece. In addition, the general topographic

2.3.4. Conclusions

Examining the torrents of each category that were studied revealed that they had similar physiographic and morphometric characteristics. This is a strong indication that the classification based on the above three variables was successful and could be a very powerful tool for land and water manager that work with torrents. Agencies and other organizations in Greece would improve their management of torrents with the adoption of this classifications scheme.

Figure 4. Different categories of torrents in Greece: a) category Oi in Mount Olympus of Macedonia (photo by G. Zaimes), b) category F in Pindos Mountain of Ipeiros (photo by N. Zaimes), c) category H in Pieria mountain of Macedonia (photo by G. Zaimes), d) category C in Seih Sou Forest in central Macedonia (photo by G. Zaimes) and e) category Ni in Phodope Mountain in Eastern Macedonia (photo by K. Vidakis).

Table 2. The 20 final categories of Greek torrents. For each category the following characteristics are described: Φ indicator, channel slope (Jm_K), watershed slope (Jm_Λ), depositional material slope ($Jm_{\alpha\pi}$), watershed area, watershed relief, the dominant geologic formation and vegetation of the watershed, the type of deposits and where these torrents are commonly found

Category	Φ	Jm_K	Jm_Λ	$Jm_{\alpha\pi}$	Watershed area	Watershed relief	Geologic formation	Vegetation	Depositional material	Region
			%		km^2	m	(symbols see Table 2)			(See Figure 2)
A	6.3-13.2	2.0-5.0	20.0-35.0	< 1.0	31-150	1000	f, m, s, c	Throughout the watershed: Shrubs. Lower watershed: cultivated areas.	Alluvial zones with only small-sized material.	Hilly areas.
B	10.0-19.0	5.0-15.0	20.0-25.0	1.0-1.5	< 15 rarely < 50	600-800	f, m, s, c	Upper watershed: shrubs with degraded forest (*Quercus* spp). Lower watershed: shrubs or cultivated lands.	Small alluvial zones with long elongated shapes.	Hilly areas.
C	11.2-18.7	5.0-10.0	20.0-35.0	1.0-2.8	21-100 rarely 101-250	800 rarely 1500	f, c, f, l	Upper watershed: *Quercus* spp forests with *Fagus sylavtica* and *Pinus nigra*. Lower watershed: cultivated lands, olive plantations and shrubs.	Alluvial with small-sized material and some larger material. Their fans have unclear boundaries	Semi-mountainous and coastal areas.

Category	Φ	Jm_K	Jm_Λ	$Jm_{\alpha\pi}$	Watershed area	Watershed relief	Geologic formation	Vegetation	Depositional material	Region
		%			km^2	m	(symbols see Table 2)			(See Figure 2)
D	15.8 - 23.0	10.0-15.0	25.0-35.0	2.0-3.0	11-50	1600-1700	f, s, f-l, n	Shrubs, *Quercus* spp, *Fagus sylavtica*, *Pinus nigra* and *Abies* ssp. Upper watershed has parts bare of vegetation.	Small-sized material and not clearly defined fans or medium to large size material with clearly defined fans.	Mountainous and semi-mountainous areas.
E	19.4 - 26.5	15.0-20.0	25.0-35.0	2.5-3.5	11-50	800 rarely 1100-1400	f, f-m, s, s-l	*Quercus* spp, *Fagus sylavtica* and *Abies cephalonica*.	Clearly defined non-conical fans.	Pindos Mountain and W. Macedonia.
F	24.5 - 29.3	20.0-25.0	30.0-35.0	3.0-3.5	1-10	300-400	f, f-m, s, s-l	*Quercus* ssp, *Fagus sylavtica*, and *Abies cephlonica*.	Clearly defined non-conical fans.	Mountain fronts of Ipeiros, Sterea Ellada and W. Macedonia.
G	8.4-15.0	2.0-5.0	35.0-45.0	0.8-1.5	51-250	1200	f, m, s, c, gn, l, pg	Upper Watershed: *Pinus* spp. and *Quercus* ssp. Lower watershed: prairies, shrubs and cultivated areas.	Wide alluvial zones with very fine-sized to sand-sized material.	Semi-mountainous areas near the coast.

Table 2. (Continued)

Category	Φ	Jm_K	Jm_Λ	$Jm_{\alpha\pi}$	Watershed area	Watershed relief	Geologic formation	Vegetation	Depositional material	Region
		%			km^2	m	(symbols see Table 2)			(See Figure 2)
H	13.2-21.2	5.0-10.0	35.0-45.0	1.5-3.0	21-250	1000-2000	s, n, n-l, c	Upper watershed: orchards, vineyards and prairies. Lower watershed: *Quercus* ssp and shrubs.	Non-conical clearly define fans with substantial percentages of large and fine-sized material.	Coastal semi-mountainous areas. Coastal mountainous areas (Peloponisos) and Mountainous areas (central Greece and Macedonia).
I	18.3-26.0	10.0 - 15.0	35.0-45.0	3.0-5.0	21-50	1300-1500	f-l	Evergreen broadleaves, *Pinus* spp and *Quercus* ssp. Bare areas in the upper watershed	For smaller Φ values the fans are clearly defined and non-conical. For larger Φ values the fans are conical.	Costal areas in northern Peloponisos and Northern Sterea Ellada
J	22.9-30.0	15.0 - 20.0	35.0-45.0	4.0-6.0	11-30	1500-2000	f-l, c-f-l, l-gn-s	Evergreen broadleaves and *Abies* spp.	Conical fans.	Mountainous areas.
K	26.5-33.3	20.0 - 25.0	35.0-45.0	5.0-8.0	11-30	1500-2000	f-l, c-f-l, l-gn-s	Evergreen broadleaves and *Abies* spp.	Clearly defined conical fans with large volumes of material.	Mountainous areas.
L	30.0-36.7	25.0 - 30.0	35.0-45.0	6.0-10.0	<10	1200	l, f, s	Evergreen broad-leaves with bare areas in the upper watershed because of the steep slopes.	Clearly defined conical fans not larger 2.0-2.5 km.	Coastal areas with steep slopes

Category	Φ	Jm_K	Jm_Λ	$Jm_{\sigma\pi}$	Watershed area	Watershed relief	Geologic formation	Vegetation	Depositional material	Region
			%		km^2	m	(symbols see Table 2)			(See Figure 2)
M	34.6-39.7	30.0 - 35.0	40.0-45.0	7.5-10.0	<10	1200	l, f, s	Evergreen broadleaves with bare areas in the upper watershed because of the steep slopes.	Not clearly defined, small in size and with steep slopes, highly influenced by external factors.	Coastal areas with steep slopes.
N i	8.9-16.6	2.0-5.0	45.0-55.0	1.0-2.0	>100	1600-1800	l, gn, g	*Pinus* spp, *Abies* spp and *Fagus* ssp.	Slopes < 2% can be quite large (5-6 km^2) and bell shaped.	Large mountains.
N ii					<100		n, sedimentary	Upper watershed: *Quercus* spp. Lower watershed: cultivated fields and *Pinus* spp.	Small alluvial zones with very fine material.	Areas with extensive clastic sedimentary material e.g. Eastern Macedonia.
O i	15.0-23.5	5.0-10.0	45.0-55.0	2.0-4.0	30-60	>1300	l, gn	Evergreen broadleaves. Upper watershed: *Pinus* spp. and *Quercus* spp.	Clearly defined conical fans. These areas are cultivated.	Mountainous areas
O ii						1000-1100	Neogenis, n		Undefined shape-wise because of the large amounts of very fine material.	Areas with extensive clastic sedimentary material e.g. Eastern Macedonia.

Table 2. (Continued)

Category	Φ	Jm_K	Jm_Λ	$Jm_{\alpha\pi}$	Watershed area	Watershed relief	Geologic formation	Vegetation	Depositional material	Region
		%			km^2	m	(symbols see Table 2)			(See Figure 2)
P	21.2 - 28.7	10.0-15.0	45.0-55.0	3.5-5.0	20-40	>1300	l, gn	Evergreen broadleaves. Upper watershed: *Pinus* spp. and *Quercus* spp.	Clearly defined conical fans with large volumes of material.	Mountainous areas.
Q	26.0 - 33.2	15.0-20.0	45.0-55.0	5.0-8.0	15-20	<1500	l, l-f-c	Highest watershed parts: bare of vegetation. Upper watershed: *Fagus sylvatika*, *Abies* spp, *Pinus nigra* and *Pinus sylvetris*. Lower watershed *Quercus* spp.	Clearly defined conical fans with large dimensions and slopes.	Mountainous and very mountainous areas.
R	30.0 - 37.1	20.0-25.0	45.0-55.0	6.0-10.0	< 10	1800-2000	l, l-f-c	Highest watershed parts: bare of vegetation. Upper watershed: *Fagus sylvatika*, *Abies* spp, *Pinus nigra* and *Pinus sylvetris*. Lower watershed: Quercus spp.	Clearly defined conical fans with large dimensions and slopes.	Mountainous and very mountainous areas.

Category	Φ	Jm_K	$Jm_Λ$	$Jm_{απ}$	Watershed area	Watershed relief	Geologic formation	Vegetation	Depositional material	Region
		%			km^2	m	(symbols see Table 2)			(See Figure 2)
S	33.5-43.8	25.0 - 35.0	45.0-55.0	7.0-13.0	3-7 rarely 10-12	< 1600 rarely 1800-1900	l, gn	*Abies* spp, *Fagus sylvatika* and *Pinus* spp.	Small sized deposits.	Mountainous and very mountainous areas.
T	28.7-45.8	15.0 - 35.0	55.0-60.0	5.0-15.0	15-20	1800	l, g	Upper watershed: *Fagus sylvatika* and *Abies* spp. Lower watershed: Prairies.	Conical fans > 4 km^2.	Very mountainous areas of Western Sterea Ellada, Ipeiros and Eastern Macedonia.

2.3. Case Study 2: Evaluating Erosion Risk for Greek Torrents Using GIS

2.3.1. Background

Soil erosion is a natural phenomenon but human activities since the beginning of the 19th century, in Europe and worldwide, have accelerated it [20, 21]. In addition, in the semi-arid and arid environments of the Mediterranean region vegetation growth and cover cen be limited substantially. The lack of extensive vegetative cover leaves large areas unprotected to intense precipitation events that also lead to accelerated soil erosion. Reducing accelerated soil erosion is essential because soil erosion can reduce soil quality that diminishes the diversity of plants, animals and microbes, ultimately threatening the stability of the entire ecosystem [22, 23].

Estimating soil erosion for torrent watersheds caused by surface runoff is a very important research field. These estimations can be: a) qualitative (intensity, degree of erosion, speed etc.) or b) quantitative (m3/year, mm/ha.year etc.). With the help of new technologies such as GIS, soil erosion estimations can be predicted remotely with high accuracy [24, 25, 26]. In order to protect watersheds for soil erosion, the detection of the areas presenting the greatest erosion potential is of particular importance. This would assist managers to prioritize their soil erosion control efforts on specific areas of the watersheds, thus reducing erosion early enough and cost-effectively.

To accomplish this with the use of GIS it is necessary to find the most important factors that influence the detachment and distribution of sediment material (soil erosion) and in general soil movement. In the literature four factors have been primarily identified [9, 27, 28]:

- Climate
- Geologic formation
- Topography, and
- Presence or absence of vegetation

The objective of this case study was to process the most important factors that influence erosion and develop maps that present the areas of greatest erosion potential, remotely. Because the climate of the watershed area of the torrent is typically very similar, only the other three factors were considered.

These three factors, geology, topography and vegetation were processed together with the help of GIS.

2.3.2. Torrent Digital Elevation Models (DEMs)

To process these three factors together, Digital Elevation Models (DEMs) were developed to represent the watershed of a torrent. The DEMs provide a better visual representation of the topography of the watershed compared to common topographic maps. The DEMs were developed with GIS tools (GRASS and MapInfo) that utilized data from topographic maps [29, 30, 31, 32, 32]. The topographic maps were provided from the Hellenic Military Geographic Service (HMGS) and had a scale of 1:50,000.

The DEM was represented in raster format that separates the watershed in many small square parts called cells and allows analyzing information for each of the cells. The cell size of our torrent watersheds differed depending on the watershed size. Smaller watershed had smaller cell sizes and larger watershed had larger cell sizes. The cell sizes ranged from 7 x 7 m to 30 x 30 m cells. The DEM allows measuring morphometric characteristics such as area, slope and channel length using data from each cell. In addition to the morphometric characteristics, the DEM allows to measure the volume of the watershed and its deposits.

The process to create the DEMs was the following: First the HMGS topographic map of each torrent was scanned and optimized in an image processing software (Adobe Photoshop). Then, these maps were geo-referenced in GRASS, by using Transverse Mercator map projection with the Hellenic Geodetic Reference System of 1987 (HGRS87). The Root Mean Square (RMS) Error of geo-referencing was at 0.5 m, an acceptable value for the scale of the maps used. Afterwards, layers were created for the watershed boundaries and elevation contour lines. Once the layers were created, the maps digitization and geo-database updating took place. Orthophotos were used to enhance the spatial accuracy of the digitized watershed boundaries. This digitization provided the appropriate data that allowed the creation of the DEMs of the torrent watersheds using the Kriging methods. The Kriging method was preferred because it best represented the topography of torrent watersheds of Greece.

DEMs were created only for 20 watersheds out of the 172 torrents surveyed for the torrent classification. The watersheds selected, best represented each category. Four of the torrent watershed DEMs are shown in Figure 5.

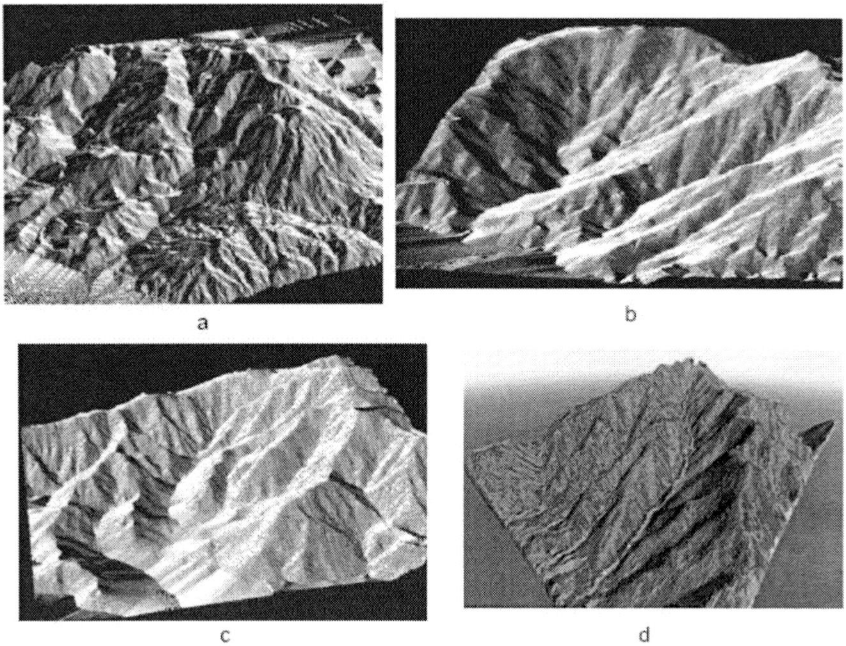

Figure 5. The Digital Elevation Models (DEMs) for some of the categories of the classification: a) category A, b) category J, c) category P and d) category T.

2.3.3. Erosion Potential 3-D Maps

For each of the 20 selected torrent watersheds, several layers with different information were created by digitizing maps. These layers were the vegetation, geology and slope of the watersheds.

Indicator values for each layer were developed to describe erosion potential. The larger the indicator value the greater the erosion potential. For example granites (very resistant to erosion) had a value of 1, the limestones 2, the paragneis 3, the flyschs 4 etc. Similarly numerical values were generated for the different vegetation types and different slopes.

The indicator values of each layer were summed for each cell providing the erosion potential of the cell as a combination of all three layers. Then depending on the cell value, it was placed in a specific erosion potential class that was assigned a specific colour. This colour was presented on the 3-D maps and indicated the erosion potential of the various parts of the watershed. In general these maps provide information on sediment movement. In addition, because the watershed provides the majority of the torrents sediment load, indirectly it describes the torrent sediment transport behavior.

The classes used to describe erosion potential on the maps and the assigned colors were the following::

- *Low Erosion Potential* (indicated by blue). The sum of the three indicator values ranged from 3-4.
- *Medium Erosion Potential* (indicated by green). The sum of the three indicator values ranged from 5-7.
- *High Erosion Potential* (indicated by yellow). The sum of the three indicator values ranged from 8-10.
- *Very High Erosion Potential* (indicated by red). The sum of the three indicator values were greater than 11.

Figure 6 presents 3-D maps from some of the torrent watersheds with the erosion potential classes.

2.4. Conclusions

The 3-D maps display the erosion potential of the different areas of the torrent watershed. This allows land and water managers to know where they should prioritize soil erosion conservation practices (areas with very high and high erosion potential) in order to minimize erosion the most. Because this can be done without in-situ measurements it can assist managers to reduce the time it would take them to survey the entire watershed. Of course an actual field survey will be required in order to verify the erosion potential of these areas.

3.0. Case Studies from Spain: Water Erosion in the La Viuda Rambla and Restoration Efforts in the Aras Torrent

3.1. Introduction

Spain is a country characterized by strong relief with a wide range of climatic conditions. These two characteristics have led to the development of two clearly distinguishable types of torrents: the *alpine torrents*, which are located in the mountain ranges (Pyrenees, Europe Peaks and Sierra Nevada) and *ramblas* (dry riverbeds or wadi), which are abundant from the coast of the

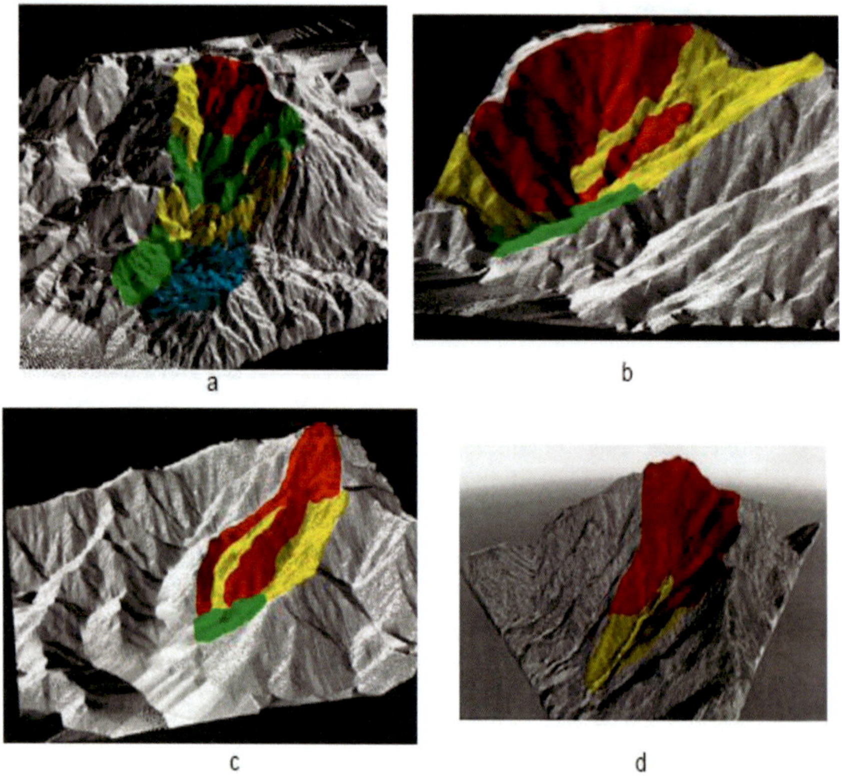

Figure 6. The Digital Elevation Models (DEM) with the erosion potential for some of the categories of the classification: a) category A, b) category J, c) category P and d) category T. There are four classes of Erosion Potential: i) Low (indicated with blue), ii) Medium (indicated with green), iii) High (indicated with yellow) and iv) Very High (indicated with red).

western Pyrenees, throughout the Levant region and the southeastern Spanish Mediterranean region (Figure 7). The word rambla comes from the Arabic, *ramla*, which means sand, making clear reference to the granular nature of its stream bed. However, in each geographical area or country it is known with a different name.

In general, it can be considered that ramblas (Figure 8a) arise from small gullies and low hills. The watershed of a rambla has one or more headwater regions, a wide gorge with a slight slope that can be several kilometres long or just a few meters. The mouth of the rambla ends either in a valley or on a plain and produces little pyramids of deposits due to its low slope. Most ramblas have water only during the rainy season. Its sediment load originates primarily from erosion, and often these loads are not massive.

Figure 7. The location of the alpine torrents, ramblas and canary ravines in Spain (from Map-of-Spain.co.uk)

On the other hand, the Spanish alpine torrents occur at high altitudes of mountain ranges with a wide catchment area and a long and deep gorge, caused primarily by the geology of the site (Figure 8b). It has constant flow throughout the year that increases only during spring and summer due to snowmelt. These torrents have steeper slopes, with an irregular torrent bed profile, banks that tend to change quickly and stream lengths that are shorter than regular streams [34, 35]. From a hydrological point of view, alpine torrents differ from regular streams because alpine torrents have:

- water flow that is faster, more variable and turbulent that carries large material (trees, gravel, boulders, etc.),
- great diversity in the way material move though the torrent (e.g. sediment loads, mass failure loads, pouring lava),
- extreme discharges that are correlated with catastrophic damages and
- continuous and abrupt transformations of their geomorphologic conditions.

In addition to the alpine torrents and ramblas, the *canary ravines* also need to be described (Figure 8c). Although the canary ravines geographically are not located in the Mediterranean region, they can be considered Mediterranean torrents because of their similarities in behavior and the damages they cause. Their watersheds have a volcanic origin and are common in the Canary Islands. The climate of these islands is influenced by the trade winds and the Azores anticyclone, creating a permanent spring climate, with intense rainfall for short periods of time, which drain through these ravines into the Atlantic Ocean.

Figure 8. There are three main torrent types in Spain: a) ramblas (photo by Robredo S. Jose C.) b) alpine torrents (photo by Garcia R. Jose L.) and c) canary ravines (Tenerife Island of the Canary Island) (photo by Garcia R. Jose L.).

While the canary ravines are an important torrent type for Spain, emphasis was given to the two true Mediterranean representative torrents. The first case study refers to the ramblas and the use of modelling to predict erosion while the second case study refers to alpine torrents and restoration efforts in order to manage their destructive behavior.

3.2. Case Study 3: Mediterranean Ramblas

3.2.1. Historical Background

The main climatic feature of the Spanish Mediterranean regions is the *cold air pool* effect. The cold air pool is a non-frontal extra tropical atmospheric disturbance, which can cause exceptionally violent and intense rainfall for a few hours or days, accompanied by lightning and hail. Its violent and intense rainfall causes large streamflow through the Spanish ramblas (Figure 9) (Table 3).

Such an effect was the cause of the flood on September 25th 1962 in Catalonia, Balearic Islands and Castellon. These heavy rains that fell in Barcelona led to the overflow of Llobregat and Besos Rivers, the latter

Figure 9. The damaging effects after a severe rainfall event in the Almeira Rambla of Spain (photo by López Cadenas F.).

Table 3. Historical flow of Spanish Ramblas

Rambla	m^3/s	Date	flow measuring site
Albuñol	2 580	1973	Albuñol, Granada
Nogalte	1974	1973	Puerto Lumbreras, Murcia
Ovejas	400	1982	Alicante, Alicante
Viuda	1500	1962	Almazora, Valencia

reaching a streamflow of 3,000 m^3/s, causing over 700 deaths and numerous damaged homes and infrastructure. The rains also affected Andratx and Palma de Mallorca, in the Balearic Islands. In Castellon, the La Viuda Rambla, a tributary of Mijares River, reached a streamflow of 1,500 m^3/s, breaking the barrier of the Maria Cristina dam and pouring water up to 1.70 m above the crest of the dam.

In October 2000, torrential heavy rains occurred again that affected Catalonia, Valencia, Murcia Regions and the provinces of Teruel, Albacete and Almeria. Rainfall amounts exceeded the 250 mm in the province of Castellon and the 500 mm in various locations in the Valencia County. The rains caused large floods in almost all rivers, streams and torrents from Tarragona to Almeria. The situation was particularly serious in the Mijares River, where again the dam of the Maria Cristina that is located on its tributary, the La Viuda Rambla, overflowed and could have broken through a crack that was created at the base of the dam. There was also extensive damage in Vinaroz, where the Cervol River overflowed, and in Morella where a flash flood occurred in the Bergantes River that also affected the Ebro River. In the Murcia Region the widespread rains affected especially Cartagena and Mar Menor, with the flooding of their ramblas causing several losses of human lives.

Finally in August 2010, a rainfall event of 200 mm/h during 2 hours occurred in the Murcia Region. The floods, as a result of the intense rainfall, caused losses of human lives, natural resources and infrastructure. The region was declared a disaster area.

3.2.2. The Case of the La Viuda Rambla

The La Viuda Rambla (translation in English *the Widow Rambla*) starts at the confluence of the Monleon River with the Carbonera Rambla, near the municipalities of Culla, Sierra Engarceran, and Useras, in the province of Castellon (Valencia Region). It flows into the Mijares River near the municipality of Almazora, just a few kilometers from the Mediterranean Sea (Figure 7).

The headwaters of the Monleon River occur at 1,300 m, forming a large canyon at its junction with the La Viuda Rambla, at the municipal districts of Benafigos and Culla (province of Castellon). Although the torrent does not usually have high streamflow most of the year, it increases substantially during fall because of heavy rainfall events and during spring due to snowpack melting and heavy rainfall events.

The watershed area of the La Viuda Rambla is about 1150 km^2, running north to south along the continental side of the coastal mountains of the province of Castellon. In general, the groundwater level of the La Viuda Rambla is lower than its own bed, so, there are significant transmission losses through the dry riverbed, year around except during the rainy season in autumn.

3.2.3. Spatial Modelling of Erosion and Deposition with USPED

The objective of this case study was to quantify the amount of sediment that can be eroded and deposited, according to the current conditions of vegetation cover, rainfall, soil and topography of the La Viuda Rambla watershed.

The USPED (Unit Stream Power - based Erosion/Deposition), is a model used for the estimation of erosion. This model can predict the spatial distribution of erosion and deposition rates for a steady state overland flow with uniform rainfall conditions [31, 33]. It assumes that the erosion transport capacity is limited. This means that water flow can transport a limited amount of sediment that is dependent on the transporting capacity of water flow. This model also assumes that the amount of sediment carried by water is always at its full transporting capacity [34]. Therefore in areas where transport capacity increases, erosion is predicted; in contrast in areas where transport capacity decreases, sedimentation is predicted [34, 37].

The USPED was used to model the distribution of erosion and deposition for the entire watershed of the La Viuda Rambla but also for its bed. The classification and delineation of land use as well as the identification of the watershed boundaries of the Rambla was done through the SIOSE (Acronym in Spanish) project. SIOSE is the Land Use Information System of Spain, which aims to integrate the information on land cover with the databases of the Autonomous Regions and the State General Administration [38]. The SIOSE is also part of the Spanish National Land Observation Plan (PNOT, Spanish acronym), which is coordinated and managed by the National Geographic

Institute of Spain (IGN, Spanish acronym) and the National Geographic Information Center (CNIG, Spanish acronym).

The results of the USPED model on the La Viuda Rambla watershed are shown in Table 4. These results show that erosion occurs in 72% of the watershed area, while in the remaining 28%, deposition takes place. On the rambla bed, deposition occurs in its majority (63%). Based on these percentages, the watershed has significantly different erosional/depositional processes compared to the rambla's bed. This was also confirmed with field observation that found large amounts of material of all sizes deposited on the rambla bed. This is a very important finding because it indicates that the majority of the deposited material originates from the watershed and not the torrent channel. Measures to reduce erosion should be taken primarily in the watershed area and not the torrent channel.

3.2.4. Conclusions

In most environmental research fields, one of the key challenges is to model physical processes with sufficient precision and efficiency. The rapid development of information technologies provides new tools and opportunities to address extremely complex environmental issues, such watershed and torrent bed erosion and deposition.

Table 4. Erosion and deposition percentages estimated with the USPED model for the entire watershed of the La Viuda Rambla and for its bed

Erosion/Deposition	Basin			riverbed		
	Area (ha)	%	%	Area (ha)	%	%
Very High Erosion (> 200 t/ha.year)	1845	1,6		42	3,1	
High Erosion (50 - 200 t/ha.year)	3512	3,1	72	59	4,4	37
Moderate Erosion (10 - 50 t/ha.year)	9526	8,3		101	7,5	
Low Erosion (0 a 10 t/ha.year)	67945	59,4		294	22,0	
Low Deposition (0 a 10 t/ha.year)	20297	17,7		311	23,2	
Moderate Deposition (10 - 50 t/ha.year)	6120	5,3	28	239	17,9	63
High Deposition (50 - 200 t/ha.year)	3302	2,9		161	12,0	
Very High Deposition (> 200 t/ha.year)	1932	1,7		133	9,9	
Total	114479	100	100	1340	100	100

The development of the new information technologies for map generation can facilitate the more accurate detection of specific areas, over large watershed, such as those that need to be protected in order to reduce soil losses caused by water flow. Integrating GIS with distributed erosion models that can incorporate deposition processes (e.g. USPED model) can improve the analysis of sediment transport on complex surfaces, avoiding the overestimation of net erosion. If these models are validated and calibrated with the appropriate field data, they can become a powerful support tool in conservation and restoration management projects for torrents throughout the Euro-Mediterranean region.

3.3. Case Study 4: Alpine Torrents

3.3.1. Historical Background

The alpine torrents are located in high elevation areas such as the Pyrenees, Sierra Nevada and Europe Peaks mountain ranges of Spain (Figure 7). The most catastrophic effects of these torrents in Spain in the past, have occurred in the headwater tributaries of the Ebro River in the provinces of Huesca and Lleida. Since the early twentieth century, significant efforts have been made by the Spanish Forest Service to restore these types of torrents. Their efforts led to the Spanish Watershed Restoration methodology that has led to effective and efficient changes in river and torrent watersheds that can still be observed today. The basis of this methodology was the empirical experiences of technicians who were specialized in the restoration of mountainous areas.

This methodology includes the use of man-made structures such as check dams that are useful in retaining sediment and other material and stabilizing unstable slopes from possible landslides. The combination of these structures, with reforestation efforts that were carried out simultaneously, has made possible to control erosion and reduce direct surface runoff [39, 40]. This dual perspective has been followed worldwide for this type of restoration efforts. In addition, the Spanish Forest Service adopted successful restoration examples from France and Central European countries such as Switzerland and Austria. These countries developed large engineering projects associated not only to torrents but also to the snow avalanches that can indirectly impact torrents.

3.3.2. The Aras Torrent Watershed

The Aras Torrent watershed is located in Tena Valley, in the Aragonese Pyrenees and administratively the entire watershed belongs to the municipality of Biescas (province of Huesca) that includes the villages of Aso, Yosa and Betes (Sobremonte Region). One of its main features are the steep slopes and water falls, due to a former glacier's activity at Tena Valley. The geological substrate is mainly sandstone and marl, partially covered by deposits of the old glacier (moraines). Typically, such deposits are very sensitive to gully erosion and landslides [41].

The section of the Pyrenees Mountains that the Aras Torrent is situated in, has peculiar physiographic and geological features. These features along with the frequent intense rainfall events led to the formation of the following alpine torrent morphology: reception basin, gorge and alluvial fan. The Valposata Peak is the tallest point of the Aras Torrent watershed with an altitude of 2,190 m and the lowest is the mouth of Gallego River, at 837 m. Aso is the name of the highest branch of the torrent while the two main tributaries of the Aras Torrent are the Betes and Yosa. The total watershed area above the alluvial fan is 19 km^2.

The most notable characteristic of the climate in the Aras Torrent watershed is the rainfall regime. The mean annual rainfall is around 1,200 mm with most of rainfall falling during winter and autumn, because of the influence of the Atlantic Ocean. It must be noted that heavy rainfalls also occur during spring while sudden intense rainfall events for short periods of time (storm events) also occur during summer.

Rangelands are the main vegetation type in the headwaters of the torrent (1,600-1,800 m). The hills and ridges are exposed to wind and species such as *Genista horrida* and *Buxus sempervirens* dominate these areas. In addition, three forest types are present in the Aras watershed: oak (*Quercus* spp) forests in the lower part of the watershed (850-1,400 m), pine (*Pinus* ssp) forests in the middle elevations (1,300-1,800 m), and riparian forests along the banks of the torrents. The watershed has no large areas with pure shrub stands. In general, shrubs are intermixed with trees, in meadows and pastures. As for the bottom parts of the watershed they are dominated by meadows and crop fields.

The land use distribution in the past was close to the desirable, in terms of naturally regulating surface runoff and controlling sheet erosion and gullies. From the beginning of the past century the Aras Torrent activity appears to be increasing and causing damages. An example is the destruction of the main road from Panticosa, Spain to France primarily because of slope instability and soil losses from areas used in forest operations and cattle grazing. This problems indicate that measures needed to be taken to restore the Aras Torrent.

3.3.3. Restoration Efforts for the Aras Torrent

3.3.3.1. Restoration Efforts: 1907-1995

The first hydrologic restoration in the Aras Torrent watershed started in 1907. These efforts continued for the next 88 years. During this period two large dams were constructed: one, in the main channel, near Aso Village and the other one in Yosa Torrent, near Yosa Village. Once the dams were completed, additional repairs were done whenever damages occurred.

In the alluvial fan, the channel bed slope was adjusted by stonework sills consisting of 32 steps in order to channel properly the water in the alluvial fan. The steps lengths ranged from 12 to 40 m and their heights from 1 to 2 m.

In the gorge section, a series of 10 check dams were built to divert water flow. The check dam heights ranged from 2 to 3.5 m, and were 12 to 26 m apart. Longitudinal walls along the torrent channels connected the check dams with each other. The length of these walls was 170 m with an elevation difference of 35 m.

In the lower and middle reaches of the gorge, 17 small check dams were built (1 m to 9 m in length) to protect the torrent bed and banks, including the stilling basin. In the upper gorge (above Betes Tributary) only two small dams (5 and 7 m in height) were built at the foot of the hillside where Yosa Village is located. These structures stabilize the hillside, near the village. In addition in the gorge, walls of masonry were built, in order to divide and decrease the slope, preventing surface erosion. Numerous barricades, small dams and embankments were constructed in the adjacent small ravines.

Finally in the Aso and Selva Tributaries, three check dams (with masonry and concrete) were built. A discharge of 125 m^3/s was used to design these three check dams because it is the highest discharge that occurs in watersheds of this size with these characteristics.

3.3.3.2. The Flood Event on August 7th, 1996.

On August 7th,1996 there was a huge storm over the Aras Torrent watershed in the Tena Valley causing a sudden and violent flash flood. According to the Spanish National Institute of Meteorology, the rainfall intensity that fell over the basin, between 3 and 9 p.m., was more than 250 mm. There were no records of such high rainfall events in the region. Still the most critical part for the flooding event was the 150 mm of rainfall that fell in

just one hour. The largest water flow contributions came from the Betes Tributary and its banks suffered important damages due to water erosion.
The high streamflows of the Aras Torrent continued towards the Gallego River, flooding its middle and lower portions, damaging most of the corrective structures built and mobilizing large quantities of material (Figure 10). The front wave of the streamflow buried the head of the gorge and afterwards spread all over the alluvial fan where the camping site of "Las Nieves" was located, causing 87 casualties and 134 injuries. The storm event also caused serious damages to the main road and surrounding infrastructures of Sobremonte Region.

Out of the 41 structures built in the main channels of the torrents, only 13 were still standing and operational after the flood event. The channel of the alluvial fan was in good condition, although there were large amounts of deposited material, especially in the upper part. The rest of the structures, like the low channel walls and small dams, held up well to the flooding event.

To estimate the maximum streamflow caused by this storm many models were used. The streamflow estimates ranged from 100 to 500 m^3/s. The reason for this wide range was the complexity of the estimation because of the

Figure 10. Check dams damaged by the stream flow and the material that were carried near the Gallego River during August 7th 1996 storm (photo by Nicolas J.).

Figure 11. The last check dam before the stonework sills in the alluvial fan of the Aras Torrent (photo by Gimenez S. M).

high rate of material transported by the flow, and the absence of actual streamflow measurements. These two factors led to a significant degree of subjectivity when applying these hydrometeorological models. Still it is most likely that the discharge did not exceed the 250 m^3/s. The reasoning is that the last check dam that withstood the flood event, (Figure 11) was constructed with a capacity of 250 m^3/s [40].

3.3.3.3. Maintenance and Restoration Efforts: 1997-Today

In order to restore the Aras Torrent, after the flood event of August 1996, the Spanish Government agreed to provide financial aid and developed a new Watershed Restoration Plan. In addition, new legislation was passed that gave priority to the hydrologic restoration of the region. The restoration efforts for each of the three main morphologic features of the torrent (reception basin, gorge and alluvial fan) are described below. These efforts were done by the public company Tragsa (Agrarian Transformation Company Inc.) in a very short period of time.

Figure 12. A series of check dams in the: a) upper part and b) middle part of the gorge of Aras Torrent (photos by Gimenez S. M).

Reception Basin: Most important efforts were done through reforestation because of the effectiveness of the previously reforested areas in reducing soil erosion that was evident after the flood event in 1996. In these reforestation efforts many hectares were re-established with endemic tree climax species of the region, such as *Pinus sylvestris* in the lower parts of the watershed and *Pinus uncinata* in the higher parts. The initial tree planting took place on the hillsides to prevent sheet erosion thus reducing the degradation of the upper soil horizons.

Gorge: In the gorge the series of check dams in the upper (Figure 12a) and middle part (Figure 12b) and the check dam in the lower parts along with longitudinal side walls were repaired (Figure 11). These check dams and side walls were damaged during the large flood event. The purpose of these structures was to stabilize the torrent bed and guide the streamwater into the main channel towards the alluvial fan of the torrent [41].

Alluvial Fan: Its reparation consisted of longitudinal structures. Specifically, the channel bed slope was adjusted by stonework sills (34 stairs). The purpose was to fix the hydraulic axis, to protect the torrent banks control the torrent bed stability and increase its transport capacity during floods

(Figure 13). In addition breakwaters, protective walls and canopy structures were constructed.

3.3.4. Conclusions

The restoration efforts described for the Aras Torrent have been successful up to a point. So applying this methodology that is known as Watershed Restoration in Spain, in the future is possible for alpine torrents but also potentially for the ramblas of the semi-arid Mediterranean. Still this methodology needs to adapt to new approaches for planning activities in watersheds to become more effective. To achieve this it needs to integrate landscape, conservation and socio-economic and other factors that have not been not been considered in the past but also utilize new tools such as the USPED model that analyzes the distribution of erosion and deposition that was presented in Case Study 3.

CONCLUSION

Torrents are the major stream type in the Euro-Mediterranean region. They play an essential role in the hydrologic cycle of the region. Water flows through their channels eventually reaching the Mediterranean Sea but also recharging groundwater aquifers through transmission losses. Torrents can also provide water for human needs, as a renewable natural resource, in this

a b

Figure 13. Adjustments of the channel bed slope by stonework sills in the alluvial fan of the Aras Torrent: a) view from the last check dam in the gorge and b) view from Gallego River (photos by Gimenez S. M).

water-scarce region. Finally, the biodiversity of the region is also dependent on torrents since their riparian areas are essential habitat for many species including many endangered and rare species.

While torrents are beneficial to the region, their extreme and sudden floods that carry large loads of sediment and other material have major negative impacts on the landscape and the adjacent communities. Every year torrents cause severe damages to cities and towns and other human infrastructures and in some cases even cause the loss of human lives.

These facts indicate that managing torrents in the Euro-Mediterranean region is a priority in order to exploit their potential benefits but also to minimize their potential damages. The development of a classification system for torrents would be a very important tool for land and water managers in order to improve torrent management. This was attempted for Greece by studying more than 170 of its torrents. The end result was a classification system with 20 categories based on the torrent's main channel slope and the watershed slope. Early indications show that this classification covers quite effectively the entire range of torrent types in Greece. The classification can provide important information on torrent behavior and help in the development of torrent watershed management plans. The simplicity of the classification makes it easy to use by managers while it also has the potential to be adapted for the entire spectrum of torrents in the Euro-Mediterranean region.

Another key element for effective torrent management is understanding its erosional process and sediment transport capacity. The capacity of torrents to transport large amounts of material is one of its most damaging characteristics. To reduce sediment transport capacity it is important to identify the major sources that produce the transported material. With the use of GIS, 3-D maps of torrent watersheds can be created indicating the erosion potential of the various areas of the watershed. These maps are created by combining spatial information on the vegetation, geology and topography of the watershed, the main factors determining the erosivity of an area. Climate was not considered as a factor because it can be consider uniform throughout the torrent watershed. By locating the areas of high erosion potential, land and water managers can prioritize erosion control efforts in these areas.

In addition, the USPED model is presented that determines sediment transport in the entire watershed but also along the torrent bed. While most models concentrate on erosion, this model also incorporates sediment deposition providing a more holistic and accurate picture of sediment transportation in the watershed and torrent bed. This provides another dimension to better understand the sediment transport dynamics of the torrent watershed. This added dimension provides managers a more accurate estimate of net erosion because it avoids the overestimation of the erosion. While this model has been primarily used for the ramblas of Spain with the appropriate field calibrations, it could be used for conservation management projects for torrents throughout the Euro-Mediterranean region.

Actual restoration efforts to reduce the negative impacts of a torrent are also demonstrated. While a lot of the efforts described included civil engineering structures such as check dams, side walls, stonework sills, reforestation was also an essential part of their success. In general to have successful restoration efforts, it is essential to understand the torrents behavior and work with nature (e.g. natural mechanisms that reduce erosion) because it will provide long-term sustainable and cost-effective solution for torrent watersheds.

Even today, throughout the Euro-Mediterranean region torrents continue to cause significant damages and are not exploited to their fullest potential. The case studies described in this chapter indicate tools that utilize new technologies to better understand torrent behavior and demonstrate ways to control torrent damages that can help land and water managers throughout the Euro-Mediterranean region improve torrent management.

ACKOWLEDGMENTS

We would like to thank the reviewers Dr. Georgios Mallinis and Dr. Fernando Garcia Robredo for their time and effort and insightful comments. Their comments were very helpful in improving our book chapter. Dr. Giorgos Mallinis is an Adjunct Lecturer in Remote Sensing and GIS at the Department of Forestry and Management of the Environment and Natural Resources, Democritus University of Thrace, Orestiada, Greece. Dr. Fernando Garcia Robredo is the Deputy Director for Curriculum Planning and Development at the School of Forestry of the Universidad Politécnica de Madrid, Spain.

REFERENCES

[1] Vita-Finzi, C. The Mediterranean Valleys: Geological Changes in Historical Times. Cambridge University Press: London, UK, 1969; pp. 232.

[2] Calbo, J. In: Water Scarcity in the Mediterranean Perspectives under Global Change. Sabater, S.; Barceló, D.; Eds. The Handbook of Environmental Chemistry, 1st Edition. Springer: London, UK and New York, NY, 2010; Vol. 8, pp.1-14.

[3] Wainwright, J.; Thornes, J. B. Environmental Issues in the Mediterranean: Processes and Perspectives from the Past and Present. Routledge: London, UK, 2004; pp. 479.

[4] Kosmas, C.; Danalatos, N. G.; Lopez-Bermudez, F.; Romero-Diaz, M. A. In: Mediterranean desertification. A mosaic of processes and responses. Geeson, N.A.; Brandt C.J.; Thornes, J.B.; Eds. John Wiley: Chichester, England, 2002; pp. 57-70.

[5] Gordon, N. D.; McMahon, T. A.; Finlayson, B. L.; Gippel, C. J.; Nathan, R. J. Stream Hydrology: An Introduction for Ecologists. John Wiley: Chichester, UK, 2004; pp. 429.

[6] Leopold, L. B.; Wolman, M. G.; Miller, J. P. Fluvial Processes in Geomorphology. Dover Publications: New York, NY, 1992; pp. 522.

[7] Bull, L. J.; Kirkby, M. J. In: Hydrology and Geomorphology of Semi-arid Channels. Bull, L.J.; Kirkby, M.J., Eds. John Wiley: Chichester, UK, 2002; pp. 1-15.

[8] Nanson, G. A.; Tooth, S.; Knighton, A. D. In: Hydrology and Geomorphology of Semi-arid Channels. Bull, L.J.; Kirkby, M.J., Eds. John Wiley: Chichester, UK, 2002; pp. 17-54.

[9] Kotoulas D. (in Greek). Ορεινή Υδρονομική Τόμος I, Aristotle University: Thessaloniki, Greece, 1998; pp. 669.

[10] Allan J. D.; Castillo, M. M. Stream Ecology: Structure and function of running waters. Springer: Dordrecht, Netherlands, 2007; pp. 338.

[11] Dimopoulos P.; Papastergiou E.; Sycora K.; Georgiadis T.,Babalonas A., Dafis S. (in Greek). Συλλογή και ανάλυση φυτοκοινωνιολογικών δεδομένων για τη βλάστηση της Ελλάδας . The Goulandris Natural History Museum. Greek Biotope-Wetland Centre. Athens, Greece. 1995; pp. 94.

[12] Kotoulas D. (in Greek). Οι χείμαρροι της Βορείου Ελλάδος. Scientific Yearbook Faculty of Geoponodasologikis (annex), Thessaloniki, Greece, 1969; vol. 13, pp. 178.

[13] Rosgen D. Applied River Morphology; 2nd edition; Wildland Hydrology: Pagosa Springs, CO, 1996; pp. 390.

[14] Emmanouloudis, D. 1990. (in Greek) Οι φυσικοί αποθετικοί σχηματισμοί των χειμαρρικών ρευμάτων στον Ελλαδικό χώρο.

Dissertation, Department of Forestry and Natural Environment, Aristotle University: Thessaloniki, Greece, 1990; pp. 300.
[15] Hampel R. 1980 (in German). Wildbach und Lawinenverbau 1980, 44, 71-102.
[16] Hook R. L.; Rohrer W. Geol. Soc. Am. Bul. 1977, 88, 177-1182.
[17] Emmanouloudis D.; Filippidis E. 1998. Proceeding of the International Conference Protection and Restoration of the Environment IV, Department of Civil Engineering, A.U.Th.: Sani, Halkidiki, Greece, 1998; Vol. II, pp.751-759.
[18] Emmanouloudis D.; Filippidis E., 1999. Integrated Methods in Catchment Hydrology -Tracer, Remote Sensing and New Hydrometric Techniques (Proceedings o f IUGG 99 Symposium HS4). IAHS Publ. no. 258; IAHS Press: Wallingford, UK, 1999; pp. 275-279.
[19] Aulitzky, H. Symposium Inteipraevent 1980, 4, 285-309.
[20] Bakkera, M. M.; Goversb, G.; Doornc, A.; Quetierd, F.; Chouvardase, D.; Rounsevell, M. Geomorphology 2008, 98, 213-226.
[21] Dotterweich, M. Geomorphology 2008, 101, 92-208.
[22] Pimentel, D.; Harvey, C.; Resosudarmo, P.; Sinclair, K.; Kurz, D.; McNair, M.; Crist, S.; Sphpritz, L.; Fitton, L.; Saffouri, R.; Blair R. Science 1995, 267, 1117-1123.
[23] Altieri, M.A.1999. Agr. Ecosyst. Environ. 1999, 74, 19-31.
[24] Jain, M.K.; Kothyari, U.C. 2000. Hydrol. Sci. J.1 45, 771-786.
[25] Fistikoglu, O.; Harmancioglu, N.B. 2002. Water Resour. Manage. 16, 447-467.
[26] Bhattarai, R.; Dutta, D., 2007. Water Resour. Manage. 21, 1635-1647.
[27] Charlton, R. Fundamentals of Fluvial Geomorphology. Routledge: New York, NY, 2008; pp. 234.
[28] Ritter, D. F.; Kochel, R. C.; Miller, J. R. Process Geomorphology. Wm. C. Brown Publishers: Dubuque, IA, 1995; pp. 544.
[29] Line, D. E.; Coffey, S. W.; Osmond, D. L. Trans. ASAE 1997, 40, 971-975.
[30] Mitas, L.; Mitasova, H.; Brown, W.; Astley, M. Proc. of the III. Int. Conf. on Integration of Environmental Modeling and GIS. National Center for Geographic Information and Analysis: Santa Barbara, CA, 1996; CD pp. 15.
[31] Neidig, C. A.; Gerdes, D.; Kos, C. GRASS 4.0 Map Digitizing Manual: v.digit. U.S. Army Corps of Engineer: Champaign, IL, 1991; pp. 53.

[32] Shapiro M.; Westerveld J., 1992. R.MAPCALC. An algebra for G.I.S. and image processing. U.S. Army Corps of Engineer: Champaign, IL, 1992; pp. 22.

[33] GRASS Development Team. Geographic Resources Analysis and Support System – GRASS. Baylor University: Waco, TX, 1999; pp. 318.

[34] Palacio, E. (in Spanish). La restauración hidrológico-forestal en España: Gestión sostenible de los recursos suelo, agua y vegetación. Ministerio de Medio Ambiente. Organismo Autónomo de Parques Nacionales: Madrid, Spain, 1999; pp. 75.

[35] Palacio, E. (in Spanish). Las ramblas: los ríos invisibles. Ministerio de Medio Ambiente. Organismo Autónomo de Parques Nacionales. Madrid, Spain, 2002; pp. 42.

[36] García Rodríguez, J. L.; Giménez Suárez, M. J. Hydr. Eng. 2010, 15, 714-717.

[37] Giménez Suárez, M. (2008) (in Spanish). Metodología de cálculo del factor topográfico, LS, integrado en los modelos RUSLE y USPED. Aplicación al Arroyo del Lugar, Guadalajara (España) (E-text type-In Spanish). Ph. D Thesis. ETSI Montes. Universidad Politécnica de Madrid. Madrid, Spain, 2008; UPM Digital Library; Available at: http://oa.upm.es/1914/

[38] García Rodríguez, J. L.; Giménez Suárez, M. (in Spanish). Proyecto PATFOR. Estudio de la pérdida de suelo por erosión hídrica en la Comunidad Valenciana. E.T.S.I. Montes. Universidad Politécnica de Madrid [Unpublished]: Madrid,, Spain, 2010; pp. 89.

[39] García Rodríguez, J. L. (in Spanish). In: La erosión hídrica y los procesos erosivos. Modelos de evaluación, en la Ingeniería en los procesos de desertificación. López Cadenas de Llano, F.; Ed.; TRAGSA/Mundi-Prensa: Madrid, Spain, 2003; vol. 1, pp 333-423.

[40] López Cadenas de Llano, F. (in Spanish). Restauración hidrológico-forestal de cuencas y control de la erosión. TRAGSA/Mundi-Prensa: Madrid, Spain, 1998; vol. 1, pp 940.

[41] Nicolás Rodríguez, J. (in Spanish). Restauración hidrológico-forestal de la cuenca del torrente Arás. TRAGSA/Mundi-Prensa: Madrid, Spain, 2001; vol. 1, pp. 221.

In: Mountain Ecosystems
Editor: Kevin E. Richards, pp. 45-78
ISBN 978-1-61209-306-2

Chapter 2

ALTERED COMMUNITY DYNAMICS IN ROCKY MOUNTAIN WHITEBARK PINE FORESTS AND THE POTENTIAL FOR ACCELERATING DECLINES

Shawn T. McKinney[1] **and Diana F. Tomback**‡[2]*

[1]National Park Service Inventory and Monitoring Division, Sierra Nevada Network, CA [2]Department of Integrative Biology, University of Colorado Denver, Denver, CO

ABSTRACT

Whitebark pine (*Pinus albicaulis*) functions as both a keystone and foundation tree species in upper subalpine and treeline forest ecosystems throughout western North America. Populations of whitebark pine are declining nearly rangewide from a combination of threats, including the invasive fungal pathogen *Cronartium ribicola* (which causes white pine blister rust), widespread population upsurges of the native mountain pine beetle (*Dendroctonus ponderosae*), altered fire regimes and advancing forest succession, with further declines expected from climate warming. *Cronartium ribicola* is geographically the most widespread threat and

---

* National Park Service Inventory and Monitoring Division, Sierra Nevada Network, P.O. Box 700-W, El Portal, CA 95318; shawn_mckinney@nps.gov

‡ Department of Integrative Biology, University of Colorado Denver, P.O. Box 173364, Denver, CO 80217; Diana.Tomback@ucdenver.edu

could greatly reduce regional populations by itself; but, in some locations, such as the Northern Rocky Mountains, interactions of several threats are leading to especially rapid declines.

Interspecific interactions drive two key ecological processes within whitebark pine communities–seed predation and seed dispersal. Whitebark pine obligately depends on Clark's nutcracker (*Nucifraga columbiana*, Family Corvidae) for dispersal of its large, wingless seeds. Prior to seed dispersal, North American pine squirrels (*Tamiasciurus hudsonicus* and *T. douglasii*, Family Sciuridae), efficient and voracious arboreal seed predators, typically cut down whitebark pine cones for storage in middens. These cones are essentially unavailable to nutcrackers, and hence their seeds rarely lead to regeneration. We present evidence that as whitebark pine declines within subalpine forests, the rate of predispersal seed predation accelerates and the probability of seed dispersal decreases, leading to reduced tree regeneration over time. We further show that the dynamics of these interspecific interactions depend on forest composition, and under certain conditions result in a positive feedback scenario that hastens the decline of whitebark pine populations. The current management goal to restore and maintain these valuable high-elevation whitebark pine communities centers primarily on the propagation of rust-resistant genotypes within populations, which is accomplished either through natural seed dispersal from resistant parent trees, or nursery production and outplanting of resistant seedlings. We discuss how our findings could be incorporated into a restoration strategy for whitebark pine.

Overview of the Problem

Whitebark pine (*Pinus albicaulis*) is an important subalpine and treeline conifer of the western U.S. and Canada. Functionally, it serves as both a keystone and foundation species for these high elevation ecosystems. Whitebark pine is obligately dependent on Clark's nutcracker (*Nucifraga columbiana*) for seed dispersal, and thus tree regeneration. In late summer, nutcrackers seek out stands of whitebark pine in order to harvest seeds. They then store these seeds from montane to treeline elevations within a diversity of terrain. Pine squirrels (*Tamiasciurus hudsonicus* and *T. douglasii*), however, are major competitors of nutcrackers for whitebark pine seeds; they cut down cones for storage in middens as a winter food supply, and the seeds are rarely placed under conditions that lead to germination and seedling survival.

Despite their remote locations at high elevations, whitebark pine populations are declining nearly rangewide from exotic disease, native pest outbreaks, altered fire regimes, and advancing succession. Climate change is also predicted to reduce whitebark pine distribution greatly in the U.S. The magnitude of current population losses and the level of anticipated declines have prompted evaluation of whitebark pine during 2010-11 for listing as a threatened or endangered species under the Endangered Species Act. Here, we show that whitebark pine declines from these various threats may, in fact, be accelerated by both dependence on nutcrackers for seed dispersal and preference by pine squirrels for its cones. Our research suggests that when cone production drops below a threshold value, seed dispersal may be greatly reduced. The decline of trees in local populations as well as reduced reproduction, loss of genetic diversity, and loss of regeneration, propels the local population into an extinction vortex, leading to extirpation. These observations have important implications for management strategies to recover whitebark pine populations.

Whitebark Pine as a Major Forest Tree in the North American West

Whitebark pine, a conifer of upper subalpine and treeline elevations, is widely distributed throughout the higher mountains of the western United States and Canada (Arno and Hoff 1990, Ogilvie 1990). The pine belongs to a holarctic group of species generally referred to as the "five-needle white pines" (family Pinaceae, genus *Pinus*, subgenus *Strobus*), which are characterized by needles growing in fascicles of five. Among the eight five-needle white pines in western North America, whitebark pine has the largest distribution, extending from 37° to 55° latitude and from 107° to 128° longitude (McCaughey and Schmidt 1990, Olgivie 1990, Tomback and Achuff 2010) (Figure 1). The western-most distribution of whitebark pine extends from the southern Sierra Nevada of California north through the Cascade Range, Blue and Wallowa Mountains, and into the Bulkley Mountains to west of Smithers in central British Columbia; the eastern distribution extends from the Wyoming and Salt River Ranges of southwestern Wyoming, and from the easternmost Wind River Range north along the continental divide of the Rocky Mountains to northeast of McBride, British Columbia. Isolated stands in southern British Columbia and northeastern Washington bridge the two distributions; and both isolated and outlying stands occur as well in the Great Basin and elsewhere rangewide (McCaughey and Schmidt 2001).

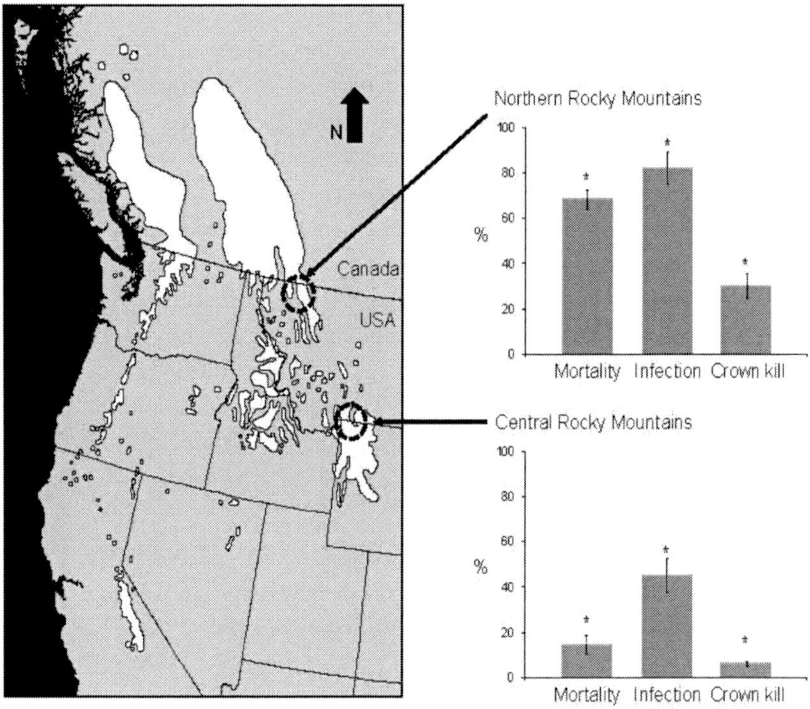

Figure 1. Distribution of whitebark pine (white) in western North America (gray) with the locations of the two study regions (black dashed circles). Data are mean values (with SE bars) for whitebark pine health parameters collected from 2004 to 2006 at multiple research sites within each region (NRM = 10, CRM = 8). Mortality is the percentage of standing trees that are dead, infection is the percentage of live trees infected with *Cronartium ribicola*, and crown kill is the percentage of a live tree's crown that is dead. Asterisks indicate significant differences (ANOVA, $P < 0.05$) between regions.

Whitebark pine grows under broad topoedaphic and climatic conditions, given its extensive geographic distribution (Arno 2001, Weaver 2001, Tomback and Achuff 2010). Consequently, it is found in diverse and distinctive forest associations (e.g., Arno 2001, Table 3a in Tomback and Achuff 2010). Whitebark pine increasingly dominates upper subalpine communities as conditions become colder and drier and where soils are well-drained; it may be outcompeted on sheltered, more mesic sites (Arno 2001, Weaver 2001). On harsh, windswept sites, whitebark pine forms climax or self-regenerating communities in which it co-dominates with other hardy conifers or forms pure stands; these community types are most common across its distribution (Arno and Hoff 1990, Arno 2001). Under less harsh and more mesic conditions, whitebark pine forms successional communities, which are renewed by fire and other disturbance (Arno 2001). Individual trees within these successional communities may persist into late seral stages (Campbell and Antos 2003). Successional whitebark pine communities are most widespread in the Central and Northern Rocky Mountains of the U.S. and Canada (Arno 2001). Furthermore, whitebark pine across its distribution forms krummholz treeline communities in the alpine-treeline ecotone (Arno and Hammerly 1984, Arno and Hoff 1990). Whitebark pine prevalence within these communities declines with snow accumulation and increases under drought and wind stress (Arno and Hammerly 1984, Resler and Tomback 2008).

Distinctive in several ways from the other five-needle white pines, whitebark pine has traditionally been included within subsection *Cembrae* of the genus *Pinus*, a small but specialized group noteworthy for sharing the unusual traits of indehiscent (non-opening) cones, large and wingless seeds, and obligate to near obligate dependence on the nutcrackers (Eurasian or Spotted nutcracker, *Nucifraga caryocatactes*; Clark's nutcracker, *N. columbiana*) for seed dispersal (Lanner 1990, Tomback and Linhart 1990, Price et al. 1998). Furthermore, whitebark pine is the only North American species of this taxon. Although recent genetic studies suggest that subsection *Cembrae* may not be phylogenetically distinct from other five-needle pine taxa (Liston 1999, Gernandt et al. 2005), seed dispersal by nutcrackers has had profound influence on the biology, distribution, ecology, and population genetic structure of the *Cembrae* pines, and this influence is best known and perhaps most highly developed in the mutualism between Clark's nutcracker and whitebark pine (Tomback and Linhart 1990, Rogers et al. 1999, Tomback 2005).

Whitebark Pine as a Keystone and Foundation Species

Keystone species are defined as having a disproportionately large influence on community diversity relative to their abundance or through their interactions with other species (e.g., Mills et al. 1993, Soulé et al. 2003). Species that provide important ecosystem functions or community stability are referred to as foundation species (Ellison et al. 2005). Whitebark pine may be considered both a keystone and foundation species in subalpine and treeline ecosystems (Tomback et al. 2001a, Tomback and Achuff 2010), thus fostering biodiversity, and providing critical ecosystem functions and community stability. Four characteristics of whitebark pine account for these ecological roles: large, nutritious seeds; seed dispersal by nutcrackers; hardy, robust seedlings; and high tolerance for dry, cold sites.

Whitebark pine fosters community biodiversity both as a wildlife food source but also as a widespread forest tree comprising many community types. The seeds of whitebark pine are the largest among conifers that grow at subalpine elevations, and so they are highly sought by a number of granivorous birds, squirrels, chipmunks, and mice, as well as by grizzly bears (*Ursus arctos*) and black bears (*U. americanus*) (Table 12-1 in Tomback and Kendall 2001, Tomback et al. 2001a, and references therein). Furthermore, there is tremendous geographic variation in community composition both in terms of forest structure and understory diversity across whitebark pines's distribution (Tomback and Kendall 2001). This results from the pine's large latitudinal and longitudinal distribution, and thus broad environmental tolerances, plus the occurrence of treeline, climax, and successional communities at different seral stages. These diverse whitebark pine communities also comprise important wildlife habitat, providing food, shelter, nest sites, and burrows to a great variety of vertebrate and invertebrate species.

Whitebark pine provides important community processes and stability, primarily due to seed dispersal by nutcrackers and the hardiness of seedlings and mature trees. Nutcrackers frequently cache whitebark pine seeds in newly burned subalpine terrain, enabling whitebark pine to establish rapidly after fire (Tomback 1986, Tomback et al. 1990, Tomback et al. 2001b). The robust seedlings are tolerant of charred seed beds and dry exposures, subsequently providing shelter and leading to the establishment of other conifers and understory species and ultimately to the successional replacement of whitebark pine. In the Rocky Mountains on particularly harsh sites, whitebark pine often acts as a "nurse" tree to spruce and fir, protecting them from high winds and ice particles (Callaway 1998). Within the extremely cold and windy alpine-treeline ecotone of the Rocky Mountain Front of Montana, whitebark pine functions as the most frequent tree to initiate krummholz tree islands. This role apparently results from the hardiness of whitebark pine seedlings combined with selection by nutcrackers of sheltered sites for seed caches, which provide some protection during early growth (Resler 2004, Resler and Tomback 2008). In general, whitebark pine grows on the harshest sites at treeline and subalpine elevations, often at the highest elevations where other conifers cannot grow (Arno and Hammerly 1984, Arno and Hoff 1990). In these upper-watersheds, the shade and shelter provided by whitebark pine canopies and tree islands protract snow melt, regulating downstream flow, and whitebark pine root systems stabilize the shallow, rocky soils, reducing erosion (Farnes 1990).

WHITEBARK PINE SEED DISPERSAL BY CLARK'S NUTCRACKER

The nutcrackers belong to the family Corvidae, and are most closely related to genus *Corvus*, the crows and ravens (Sibley and Ahlquist 1985). Both nutcracker species possess a suite of specialized morphological and behavioral traits adapting them to an annual cycle based on consuming fresh and stored tree nuts and conifer seeds. The key adaptations include a relatively long, sturdy beak that enables them to break into closed conifer cones, crack or peck open seeds, and make and retrieve seed caches (e.g., Tomback 1978, Mattes 1978); a highly developed spatial memory that enables the birds to relocate tens of thousands of seed caches each year (e.g., Vander Wall 1982, Kamil and Balda 1985); and a sublingual (throat) pouch that enables the birds to transport whitebark pine seeds from cones to dependent young or cache sites (Bock et al. 1973, Vander Wall and Balda 1977, Mattes 1978, Tomback 1978).

As noted above, whitebark pine and its subsection *Cembrae* relatives share derived traits that appear to facilitate seed dispersal by nutcrackers (Price et al. 1998). The large seeds enable the growth of robust seedlings, well-adapted to the harsh upper elevation conditions (Tomback and Linhart 1990, McCaughey and Tomback 2001); seeds in indehiscent cones cannot be dispersed by wind and require release by nutcrackers, and thus the high energy and nutrient content of seeds within cones provide rich rewards for nutcrackers seeking out cone-producing trees (Tomback and Linhart 1990, Lanner and Gilbert 1994). Furthermore, all *Cembrae* pines (except for the genetically krummholz Japanese stone pine, *Pinus pumila*) as well as other nutcracker dispersed relatives assume a flat-topped, shrubby growth form, produced by profuse branching. The branches grow vertically with horizontally-directed cones in whorls at their tips, which may render the cones highly visible to nutcrackers flying overhead, provide a stable platform for nutcrackers harvesting seeds, and enable nutcrackers to access seeds from several sides of each cone, sometimes with the support of the branch tip (Tomback, 1978, Lanner 1982, Tomback and Linhart 1990). Furthermore, studies of both whitebark pine and Siberian stone pine (*Pinus sibirica*, Subsection *Cembrae*) seeds during pre-germination treatments indicated major differences compared to other conifer seeds in morphology and storage compounds, which appear adaptive for maintaining seed viability within a soil seed bank (Tillman-Sutela et al. 2008).

Detailed overviews of the coevolved, mutualistic interaction between whitebark pine and Clark's nutcracker are presented in Tomback (2001, 2005); the important aspects are summarized here. As whitebark pine cones ripen, nutcrackers begin to harvest seeds as early as mid-July to feed themselves and their fledged but still dependent young (Tomback 1978). At this time, seed coats are light colored and soft, and nutcrackers remove only seed fragments. Thus, prior to seed ripening, nutcrackers act as seed predators rather than dispersers. Usually by mid- to late-August cone scales separate slightly and turn more brittle, seed coats harden and turn medium brown in color, and entire seeds may be removed by nutcrackers (Tomback 1978, McCaughey and Tomback 2001). With these changes, nutcracker foraging efficiency increases, and nutcrackers begin to harvest and cache quantities of whitebark pine seeds (Tomback 1978, Hutchins and Lanner 1982) (Figure 2).

Nutcrackers collect and transport harvested conifer seeds within their sublingual pouch, which may hold more than 100 whitebark pine seeds (Tomback 1978, Hutchins and Lanner 1982). They occasionally store whitebark pine seeds within a few meters to several hundred meters from source trees; they frequently fly to steep, south-facing communal storage slopes, which are often within 2 to 4 km of source trees and may accumulate little snow pack; they sometimes fly 12 km or farther from source trees to lower elevations, where whitebark pine does not grow; and they may transport seeds to treeline and alpine tundra (Tomback 1978, Hutchins and Lanner 1982, Tomback 1986, Baud 1993). Clark's nutcrackers have been reported to transport piñon pine seeds as far as 22 km for caching (Vander Wall and Balda 1977). Lorenz and Sullivan (2009), using radio-telemetry, documented nutcrackers transporting whitebark pine seeds an average of 10.6 km in the Cascade Range and a maximum of about 29 km to cache sites within the home ranges of nutcrackers.

Nutcrackers place whitebark pine seeds in caches of 1 to 15 or more seeds, with means ranging from 3 to 5 seeds per cache (Tomback 1978, 1982, Hutchins and Lanner 1982). The seeds are often buried under 1 to 3 cm of substrate, such as mineral soil, gravel, pumice, or forest litter. The caches are placed next to trees, rocks, plants, logs, and other objects, under closed and open canopy forest, and in open terrain. They are also placed at treeline among krummholz tree islands, in recent clearcuts and burned soil soon after fire, and above ground in cracks, holes, fissures, and bark of trees and logs (Tomback 1978, Hutchins and Lanner 1982, Tomback 1986, Tomback et al. 2001a, W.W. McCaughey, personal communication).

Whitebark pine seed caches are retrieved as winter, spring, and even summer food by nutcrackers to feed themselves and their young (Figure 2). Nutcrackers have been observed uncovering caches up to nearly a year after the previous whitebark pine cone crop (Tomback 1978, Vander Wall and Hutchins 1983). Typically from June through early September, buried seeds are stimulated to germinate by snowmelt and summer precipitation (Tomback 1982, McCaughey 1990). Because seeds are frequently cached in groups, seedlings often arise in clusters. Seedlings originating in nutcracker caches are the primary means of whitebark pine regeneration.

Figure 2. Clark's nutcracker (*Nucifraga columbiana*) harvesting a whitebark pine seed in the Wind River Range, Wyoming (left photo; Diana F. Tomback, University of Colorado Denver); and, a nutcracker retrieving a cached whitebark pine seed in the Bitterroot Mountains, Montana (right photo; Charles H. Janson, University of Montana).

How Seed Dispersal by Nutcrackers Influences Whitebark Pine Communities

Seed dispersal by Clark's nutcrackers has shaped tree growth form, population genetic structure, and local and regional distributions of whitebark pine. These influences result from nutcrackers caching seeds in small clusters, transporting seeds from tens of meters to tens of kilometers from source trees to cache sites, distributing caches within a variety of terrain, and storing caches both above and below the current elevational distribution of whitebark pine (Tomback and Linhart 1990, Tomback 2001, Bruederle et al. 2001, Tomback 2005).

By placing seeds in multi-seed caches, nutcrackers affect the morphology of a high proportion of whitebark pine trees. Seeds within these caches tend to germinate during the same summer or, less frequently, over several years, resulting in a group of similarly aged seedlings (Tomback et al. 1993, Tomback et al. 2001a). These seedlings appear to tolerate competition and limited space as they grow into mature trees, producing what is termed a "tree cluster" growth form, although the numbers of individuals may decline with time (Linhart and Tomback 1985, Tomback and Linhart 1990, Tomback et al. 2001b). In this growth form, the stems are contiguous or fused at the base, and the tree appears to be a single individual with multiple trunks. Genetic analyses have confirmed that the stems comprising these growth forms have different genotypes (Linhart and Tomback 1985, Furnier et al. 1987, Rogers et al. 1999).

Whitebark pine tree clusters represent a highly clumped population dispersion pattern–one aspect of a fine-scale population genetic structure produced by nutcracker seed dispersal. Further unique structure comes from nutcrackers harvesting many seeds from the same parent tree, resulting in caches and thus tree clusters comprised of genetically related individuals. But, within a given area used by nutcrackers for seed caching, and particularly in communal storage slopes, nutcrackers will cache seeds from different source trees at random; thus, neighboring caches are less likely to be genetically related. Genetic analyses have shown that stems within the same cluster are more closely related to each other than to stems within neighboring clusters (Furnier et al. 1987, Tomback 1988, Rogers et al. 1999). This unique fine-scale genetic structure is a signature of seed dispersal by nutcrackers, now known for several pines (Bruederle et al. 2001).

At the local and landscape scale, where Clark's nutcrackers cache seeds–coupled with the environmental tolerances of seedlings–determine where whitebark pine will grow (Tomback 2001, Weaver 2001). Nutcrackers store seeds throughout subalpine elevations across a diversity of topography, forest types, substrates, and conditions, as well as at and above treeline, and also store seeds below the subalpine zone (Tomback 1978, 1986, Baud 1993). Consequently, whitebark pine's elevational distribution can change rapidly in response to climate warming or cooling. In successional whitebark pine communities, seed dispersal by nutcrackers following fire leads to early regeneration of whitebark pine (Tomback et al. 1993, Tomback et al. 2001a). With respect to large-scale wildfires at subalpine elevations, which are relatively common in the central and northern Rocky Mountains, long-distance seed transportation by nutcrackers provides whitebark pine with two advantages: nutcrackers disperse whitebark pine seeds farther than most wind-dispersed seeds travel, and nutcrackers disperse seeds against prevailing winds, which otherwise may limit the distance and direction that wind-dispersed conifer seeds travel (Tomback et al. 1990). Under these conditions, whitebark pine often has an early-successional advantage over competing conifers.

At a larger geographic scale, Rogers et al. (1999) found a lack of genetic structure among whitebark pine populations in three adjacent watersheds, which they explained by long-distance seed dispersal by nutcrackers. Seed transportation distances of 12 to 29 km are more than adequate to reach different drainages on a single caching trip. At a regional and even broader spatial scale, allozyme studies suggest that whitebark pine has lower levels of genetic differentiation among populations than do most other pines, although some geographic variation does occur (see Bruederle et al. 2001 and references therein). Most genetic variation in whitebark pine results from individual differences within populations, which again may reflect the homogenizing influence of long distance seed dispersal by nutcrackers. However, whitebark pine populations do show genetic structure with respect to mitochondrial DNA haplotypes, which is the result of post-Pleistocene tree migration, via nutcracker seed dispersal, out of three different glacial refugia (Richardson et al. 2002).

SEED PREDATOR DYNAMICS

North American and Eurasian arboreal squirrels (*Tamiasciurus* and *Sciurus* genera) and modern members of the genus *Pinus* have a long-standing antagonistic, co-evolutionary relationship dating back to at least the late

Pliocene (ca. 1.7 Ma) (e.g., Hafner 1984, Axelrod 1986). Arboreal squirrels can exert significant ecological and evolutionary effects on seed traits, tree population dynamics, and forest structure by directly influencing seed fate (Steele et al. 2005). The coevolutionary interactions between the squirrel and pine taxa, with tree squirrels evolving adaptations for specialized feeding on pine seeds, and *Pinus* evolving cone and seed morphologies to inhibit seed predation, are well documented (e.g., Smith 1970, Elliott 1974, Benkman et al. 1984).

Whitebark pine cones are harvested by the North American red squirrel in the Rocky Mountain Region and Douglas's squirrel in the Pacific West Region (e.g., the Cascade Range and Sierra Nevada). Red squirrels are highly territorial central-place foragers (Elliott 1988) capable of harvesting 100% of *Pinus* cone crops (Flyger and Gates 1982). Territories are maintained by a single adult that cuts conifer cones and stores them in middens (Smith 1981). Middens are accumulations of cone debris that cover the ground to the exclusion of all living plants, and can be as large as 7 m across and 0.5 m deep (Finley 1969). Mean territory density (territories per hectare) of red squirrels in Rocky Mountain coniferous forests are generally similar, with some variation due to forest structure and composition: 0.41 in AB, Canada (Rusch and Reeder 1978); 0.72 in WY, US (Mattson and Reinhart 1997); 1.11 in WY and MT, US (McKinney and Fiedler 2010); 1.30 in CO, US (Gurnell 1984); and 1.53 in AB, Canada (Larsen and Boutin 1994).

Red squirrels preferentially select the tree species with the highest cone energy content (i.e., the energy available to squirrels in the seed endosperm and embryos of cones = no. of seeds cone^{-1} x calories seed^{-1}) (Smith 1970), and may prefer whitebark pine in Rocky Mountain upper-subalpine forests because of the pine's relatively high cone energy content (e.g., whitebark pine = 27.7 Kcal cone^{-1}, next highest subalpine fir (*Abies lasiocarpa*) = 15.7 kcal cone^{-1}) (Tomback 1982, Smith 1970). In a mixed-species Rocky Mountain subalpine forest, for example, red squirrels harvested whitebark pine cones first before moving on to the cones of other conifer species (Hutchins and Lanner 1982). Red squirrels are believed to be the major predispersal seed predator of whitebark pine in the Rocky Mountains, taking as much as 80% of the cone crop and greatly diminishing the number of seeds available for nutcracker dispersal (Hutchins and Lanner 1982, McKinney and Tomback 2007). Nutcrackers compete with red squirrels for whitebark pine seeds in habitats where red squirrels are present (Tomback 1978, Hutchins and Lanner 1982, McKinney and Tomback 2007, Siepielski and Benkman 2007a); and, where red squirrels outcompete nutcrackers for seeds, the evolution of cone and seed traits that facilitate seed dispersal by nutcrackers is thought to be constrained (Siepielski and Benkman 2007b). Red squirrel seed predation ultimately limits whitebark pine regeneration potential, because cone abundance at the time of seed ripening (late-August to early-September) significantly influences the probability of nutcracker seed dispersal (McKinney and Tomback 2007, McKinney et al. 2009): with fewer cones, the likelihood of seed dispersal declines. Therefore, the degree to which interactions among nutcrackers, red squirrels, and whitebark pine are altered regionally by mortality from blister rust and pine beetle will likely determine whether whitebark pine will self-regenerate or whether management intervention will be needed.

WHY IS WHITEBARK PINE DECLINING?

Whitebark pine is undergoing a 12-month status review by the U.S. Fish and Wildlife Service under the Endangered Species Act, following a 90-day decision published on July 20, 2010, stating that "substantial scientific..." evidence warranted further evaluation (DOI Fish and Wildlife Service 2010). It is difficult to understand how a widely-distributed conifer that occurs primarily on public lands at high elevations and in remote locations could decline to the point that triggers evaluation for threatened or endangered status. However, the factors involved in the decline of whitebark pine are primarily anthropogenic in origin and interactive, complicated by the pine's unique ecology, and highly reflective of the challenges faced by natural communities in today's changing world.

The most widespread threat is the invasive fungal pathogen *Cronartium ribicola*, which was inadvertently introduced to the Pacific Northwest by 1910 (McDonald and Hoff 2001, Geils et al. 2010). This pathogen causes the disease white pine blister rust in five-needle white pines, but requires alternate hosts to complete its life cycle. The cool, humid weather of the Northwest, abundance of white pine hosts and alternate hosts--notably native and cultivated species of currants and gooseberries (*Ribe*s spp.)--led to the rapid spread of the disease. *Cronartium ribicola* damages pines when sporulating cankers disrupt the phloem and cambium. When infections start in the canopy, the girdling process kills cone-bearing branches and weakens trees by reducing photosynthetic biomass; if infection occurs within a stem, the tree itself eventually dies. Blister rust has infected whitebark pine communities throughout their distribution in both the U. S. and Canada, with the exception of some interior Great Basin ranges (Tomback and Achuff 2010, Schwandt et al. 2010). The highest blister rust infection levels are in the northern U.S. and southern Canadian Rocky Mountains, and particularly in the Northern Divide Ecosystem (e.g., Kendall and Keane 2001, Smith et al. 2008).

Mountain pine beetle outbreaks are episodic, natural disturbances in western forests that result in forest openings and the initiation of successional communities (Arno and Hoff 1990, Schwandt et al. 2010 and references therein). The most important hosts for these outbreaks have been lodgepole (*Pinus contorta*) and ponderosa (*Pinus ponderosa*) pine, which comprise major forest types throughout the West. However, during peak outbreaks, mountain pine beetles have moved into higher elevation white pine forest communities. Outbreaks in whitebark pine forests have been dated to the 17th century (Perkins and Swetnam 1996), and the outbreaks between 1909 and 1940 and from the 1970s to the 1980s killed tremendous numbers of whitebark pine, resulting in the still-standing "ghost forests" of the central and northern U.S. Rocky Mountains (Perkins and Swetnam 1996, Kendall and Keane 2001, Logan and Powell 2001). Beginning in the late 1990s, mountain pine beetle outbreaks again moved into whitebark pine forests throughout the western U.S. but also in western Canada, achieving an unprecedented geographic scale and level of whitebark pine mortality with no sign of downturn (Taylor and Carroll 2004, Gibson et al. 2008). The extent and intensity of the current outbreaks are explained both by the large expanses of mature lower elevation pines and by climate warming. Warmer temperatures have facilitated beetle population growth and caused drought-stress, which reduces whitebark pine defenses (Logan and Powell 2001, Logan et al. 2003). The magnitude of whitebark pine losses in the Greater Yellowstone Ecosystem, in particular, has been considered historically unprecedented, and a threat to the persistence of functional whitebark pine communities (Logan et al. 2010).

Fire suppression and advancing succession have further reduced the areal extent of whitebark pine on the landscape. Fire suppression in the western forests of the U.S. dates to the settlement period of 1850 to 1904, with more aggressive suppression after 1905 (Arno and Allison-Bunnell 2002). In western Canada, suppression efforts began in the early to mid-20th century (Taylor and Carroll 2004). Suppression has altered historical fire regimes and decreased the area burned each year (e.g., Keane et al. 2002). For whitebark pine, a number of studies in the northern Rocky Mountains demonstrate a reduction in fire frequency and consequent loss of whitebark pine from seral communities, although these processes are not evident in all regions (Tomback and Achuff 2010).

Whitebark pine is one of the most vulnerable forest species to climate change, because of its high elevation distribution. Several studies have specifically included whitebark pine in their bioclimatic models, essentially reaching similar conclusions (Hamann and Wang 2006, McKenny et al. 2007, Warwell et al. 2007, Schrag et al. 2008). Assuming unimpeded dispersal, whitebark pine would lose much of its U.S. distribution by the end of the 21st century, but its projected distribution would shift further north. These niche-based models must be regarded as coarse-scale predictions that overestimate distributional shifts, since topographic variation with regions and ecological processes are not considered (e.g., Morin and Thuiller 2009). The challenges will be to manage healthy populations of whitebark pine so that nutcracker populations visit these high-elevation forest communities and whitebark pine seed production is sufficient to enable the elevational, topographic, and latitudinal distribution shifts required with changing conditions.

These threats to the survival of whitebark pine must be strategically managed in order to maintain critical population sizes. However, all of the above challenges to whitebark pine—blister rust, mountain pine beetle, fire regimes, and climate change—are interacting, with unclear outcomes (Tomback and Achuff 2010). Here we show that the historical, co-evolved interactions between whitebark pine and nutcrackers and pine squirrels are altered as whitebark pine populations diminish, resulting in highly accelerated losses.

Examining Seed Predation and Dispersal in Relation to Whitebark Pine Forest Health and Composition

Study Area

We conducted our research from 2004-2006 in the Northern and Central Rocky Mountains, USA. The two regions were selected because we expected whitebark pine forests in each to be distinctly different in the incidence and severity of blister rust infection, successional trends, and in whitebark pine tree mortality from blister rust, mountain pine beetle, and other factors, based on our previous investigations (Tomback et al. 2001a, McKinney and Tomback 2007, McKinney et al. 2009, McKinney and Fiedler 2010). Our main objective was to investigate how red squirrel predispersal seed predation and Clark's nutcracker seed dispersal behavior vary under different whitebark pine health and habitat conditions, and how this variation ultimately influences regeneration. Therefore, it was critical to have two study areas that differed greatly both in whitebark pine forest structure and composition as well as in the prevalence of damage and mortality caused by white pine blister rust.

Research in the Northern Rocky Mountains (NRM) took place in Glacier National Park and the Flathead National Forest (48.8 to 48.3°N, 113.3 to 114.4°W) in northwestern Montana (Figure 1). Elevation of research sites in the NRM ranged from 1,806 m to 2,181 m above sea level. In the Central Rocky Mountains (CRM) of southwestern Montana and northwestern Wyoming, research was conducted within Yellowstone National Park and the Gallatin and Shoshone National Forests (45.1 to 44.8°N, 109.5 to 110.6°W) (Figure 1). Elevation of CRM sites ranged from 2,546 m to 2,978 m above sea level. Forest communities selected for this study were highly variable in composition and structure, even within each region, and included whitebark pine, lodgepole pine, subalpine fir, Engelmann spruce (*Picea engelmannii*), and Douglas-fir (*Pseudotsuga menziesii*). Each research site was selected based on the presence of cone-bearing whitebark pine trees, but the different research sites within each region were intentionally selected to capture variation in tree species composition and structure and in the elevational range of whitebark pine. We established research sites (NRM n = 10 sites, CRM n = 8 sites) by delineating rectangular boundaries that were 100 m wide by ≥ 200 m long within contiguous forest stands. Sites were subdivided into 1-ha squares (100 m x 100 m) to enable us to distribute our sampling efforts more equitably. Detailed descriptions of sampling design and field methods can be found in McKinney et al. (2009) and McKinney and Fiedler (2010).

Forest Structure, Composition, and Health

We documented tree mortality, blister rust infection, and crown kill in 4,496 whitebark pine trees (NRM = 2,404; CRM = 2,092) and found whitebark pine health conditions to be vastly different between the two regions (Figure 1) . Population mean vectors for the three health parameters differed significantly between the two study regions (MANOVA Wilks' λ $F_{3,14}$ = 9.806, P = 0.001), with pair-wise comparisons confirming greater mortality ($F_{1,16}$ = 12.663, P = 0.003), infection ($F_{1,16}$ = 10.403, P = 0.005), and crown kill ($F_{1,16}$ = 14.633, P = 0.001) in NRM forests (Figure 1). The poorer health conditions in NRM forests were associated with significantly lower absolute and proportional amounts of live whitebark pine. Northern Rocky Mountain forests had significantly less live whitebark pine basal area (m^2 ha^{-1}, NRM = 1.9, CRM = 14.5, $t_{0.05,16}$ = 2.747, P = 0.007), and significantly lower live whitebark pine abundance relative to other conifer species (% relative abundance, NRM = 19.4, CRM = 53.6, $t_{0.05,16}$ = 5.020, $P < 0.001$). In addition, there was significantly lower live whitebark pine tree density by diameter class (trees ha^{-1}, NRM = 7, CRM = 25, paired t-test $t_{0.05,10}$ = 4.544, P = 0.001) (Figure 3). However, it appears that the two regions were once similar in whitebark pine abundance, as the density of all standing whitebark pine trees (i.e., the number of live plus dead trees ha^{-1}) did not differ significantly by diameter class (live plus dead trees ha^{-1}, NRM = 28, CRM = 32, paired t-test $t_{0.05,10}$ = 0.459, P = 0.656) (Figure 3). There was a high correlation between the amount of live whitebark pine basal area (ln m^2 ha^{-1}) and mean annual cone production (Pearson's correlation r = 0.81, $P < 0.01$). Thus, the current depauperate conditions in the NRM resulted in significantly fewer cones produced compared to the CRM (cones ha^{-1}, NRM = 577, CRM = 3,886, $t_{0.05,12}$ = 4.743, P = 0.005).

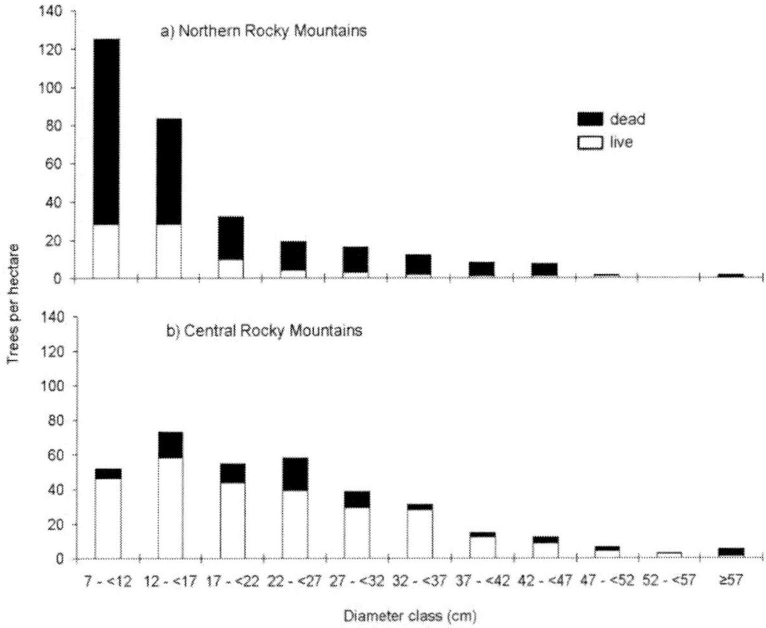

Figure 3. Size class (diameter at 1.37 m height) distributions of live and dead whitebark pine trees at study sites within two Rocky Mountain regions, USA.

Red Squirrel Seed Predation and Nutcracker Seed Dispersal

How do the large differences in whitebark pine forest conditions between the two regions affect habitat use by two key vertebrate species? Will the far fewer cones, along with the corresponding lower seed energy per hectare of forest in the NRM, influence the magnitude of red squirrel predispersal seed predation and nutcracker seed dispersal? If so, how will these altered interactions affect whitebark pine seedling regeneration? To answer these questions, we documented the degree of red squirrel predispersal seed predation, the incidence of nutcracker seed dispersal, and density of seedling regeneration within each research site. We randomly selected multiple cone-bearing whitebark pine trees in each research site (cone trees per site: minimum = 4, maximum = 16, mean = 10.6) and followed the fate of each tree's ovulate cones. Selected cone trees were more than 25 m apart with a minimum of 1 and maximum of 4 sampled trees ha^{-1}. We used tripod-mounted spotting scopes with 10–60x zoom eyepieces and handheld tally devices to count cones. Two to three observation points with unobstructed views of a tree's canopy were used to census cones on each tree. We counted the initial number of cones between 29 June and 15 July of each year, and then returned to the same observation points between 19 August and 4 September of each year and counted the remaining cones.

Whitebark pine cones are indehiscent at maturity and rarely fall to the ground during ripening without vertebrate assistance. As discussed earlier, nutcrackers typically extract seeds with the cone attached to the branch, leaving a characteristic dished-out cone after seed harvesting, and rarely dislodge cones from branches (Tomback 1978, Hutchins and Lanner 1982, Tomback 1998). In contrast, red squirrels use their sharp teeth and strong temporal muscles either to cut branch tips and drop individual or whorls of cones, or to cut the base of a single cone and eat the seeds *in situ* (Smith 1970, Hutchins and Lanner 1982). Because red squirrels were the only animals in the study area capable of cutting cones from the canopy, we attributed the difference between the initial and final cone numbers to squirrel predation (McKinney and Tomback 2007). Percent predispersal cone predation for each site was calculated annually and overall (i.e., within a year and all years combined) as the sum of cones lost to squirrels (by year, all years) divided by the sum of initial cone counts (by year, all years), multiplied by 100.

Red squirrel predispersal cone predation varied with changing whitebark pine abundance. As whitebark pine decreased in relative stand composition, the percentage of the whitebark pine cone crop taken by squirrels increased significantly (Figure 4). In forests where whitebark pine abundance is declining, proportionally more cones are depredated by squirrels prior to the onset of nutcracker seed dispersal. Because of high mortality, the majority of NRM sites were characterized by low whitebark pine abundance, and therefore, high rates of cone predation (Figure 4). Indeed, the fewer live whitebark pine trees in the NRM received greater mean predispersal cone predation–more than four times higher–than CRM trees (Table 1). Low cone production and high rates of predispersal cone predation in Northern Rocky Mountain forests resulted in few cones (and thus seeds) available for nutcrackers at the time of seed dispersal. Northern Rocky Mountain sites had less than 10% of the cone density of Central Rocky Mountain sites by the time of seed dispersal (Table 1).

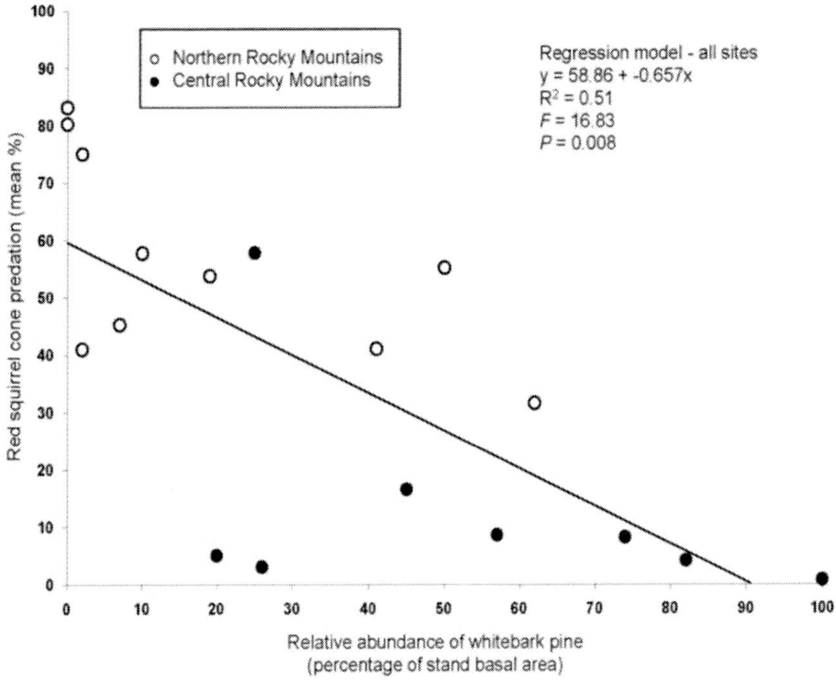

Figure 4. Simple linear regression analysis of the relationship between whitebark pine relative abundance and red squirrel predispersal cone predation (percentage of initial cone crop lost) at 18 subalpine forest sites in the Northern and Central Rocky Mountains, USA.

Table 1. Whitebark pine cone predation, cone availability, seed dispersal, and regeneration in two Rocky Mountain regions, USA

Rocky Mountain region (no. of sites)	Red squirrel predispersal cone predation (% of initial crop)	Cone density at time of seed dispersal (no./ha)	Nutcracker seed dispersal (% of sites)	Regeneration density (no. seedlings < 50 cm ht/ha)
Northern (10)	56 (6)	282 (93)	20	70 (21)
Central (8)	13 (7)	3163 (566)	100	282 (81)

Notes: Data are means with SE in parentheses, except for seed dispersal values. A site was considered to have seeds dispersed if at least one Clark's nutcracker was observed with a bulging sublingual pouch or placing seeds in its sublingual pouch.

The frequency of Clark's nutcracker occurrence (proportion of observation hours with at least one bird detected) was highly correlated with whitebark pine mean annual cone production (Pearson's correlation $r = 0.87$, $P < 0.01$) and live whitebark pine basal area (Pearson's correlation $r = 0.79$, $P < 0.01$). In the Northern Rocky Mountains, only 14% of observation hours had at least one nutcracker detected compared to 96% of observation hours in the Central Rocky Mountains. The average number of birds detected per hour was also far lower in the NRM: 0.3 compared to 6.1 in the CRM. Therefore, both the frequency of occurrence and abundance of nutcrackers declined with the poorer health conditions in Northern Rocky Mountain forests.

Nutcracker seed dispersal activity was also sensitive to the number of available cones at the time of dispersal, with far fewer Northern Rocky Mountain sites exhibiting seed dispersal activity (Table 1). Since live whitebark pine basal area was strongly associated with cone production, basal area was also a good indicator of the likelihood of nutcracker seed dispersal. Research sites where nutcracker seed dispersal was never observed had a live whitebark pine basal area of 1.58 ± 0.78 m^2 ha^{-1} (mean ± SE); sites with nutcracker seed dispersal observed in some years (and not in others) had a live whitebark pine basal area of 5.03 ± 1.01 m^2 ha^{-1}; while sites with observations of nutcracker seed dispersal in all years had a live whitebark pine basal area of 15.27 ± 2.51 m^2 ha^{-1}. Therefore, forests with live whitebark pine basal area > 5.0 m^2 ha^{-1} were able to produce enough cones, at least in some years, to attract and maintain nutcrackers through the critical period of seed dispersal.

Ultimately, lower live tree abundance, lower cone availability, and reduced nutcracker seed dispersal activity were associated with lower densities of whitebark pine regeneration (Table 1). The Northern Rocky Mountains had 75% fewer seedlings per unit area than the Central Rocky Mountains (Table 1).

Conclusion and Management Recommendations

Our findings show that whitebark pine mortality can lead to accelerated population declines over time through changes in both mutualistic and antagonistic relationships. Cone production declines resulting from increasing crown kill and tree mortality are further exacerbated by red squirrel cone predation and directly constrain the probability of nutcracker seed dispersal, and therefore, seedling regeneration. This creates a positive feedback scenario where fewer trees are recruited into the cone-producing age-class, resulting in a continual decline in cone production capacity and still fewer seeds dispersed (Figure 5). The long-term prognosis for whitebark pine communities that have entered this cycle is bleak, calling for direct management intervention.

Tomback and Kendall (2001) pointed out that without rapid and large-scale management intervention, the combination of the white pine blister rust epidemic, advancing succession, and mountain pine beetle outbreaks will result in decimation of whitebark pine throughout southwestern Canada and the northwestern U.S., and particularly in the U.S. NRM. They suggested that with on-going population losses, whitebark pine's role as a keystone (and foundation) species will diminish within communities, resulting in a loss of biodiversity and ecosystem function. Since the publication of Tomback and Kendall (2001), whitebark pine populations have continued to decline with the spread and intensification of blister rust, and with major losses from mountain pine beetle. Concerns about functional losses in the Greater Yellowstone Ecosystem have been fueled by unprecedented mountain pine beetle mortality in whitebark pine forests plus increasing blister rust infection levels in that region (Logan et al. 2010, Tomback and Achuff 2010).

In this chapter, we demonstrate that declines in nutcracker seed dispersal are occurring in many whitebark pine forests in the NRM, likely contributing to the low levels of whitebark pine regeneration, as predicted by Tomback and Kendall (2001). Here, we repeat the cautionary message from Tomback and Kendall (2001): The decline of whitebark pine may accelerate as populations become smaller, as the result of positive feedback effects. The positive feedback comes from four processes—the four "extinction vortices" of Gilpin and Soulé (1986): demographic variation, population fragmentation, inbreeding depression, and genetic drift. These processes result from reduced population size and act to further reduce population size and especially the effective population size N_e of a species, on a trajectory leading to extinction. We have documented reduced seed availability, reduced seed dispersal by nutcrackers, and lower whitebark pine population size, which in turn contribute to both demographic variation and fragmentation. With few reproductive whitebark pine trees remaining within some forest stands, self-pollination and other degrees of inbreeding become more likely, as well as reduced genetic variation resulting in genetic drift. Genetic drift reduces the potential for adaptation in local populations. Any one of these processes, or several in combination, may lead to local or even regional extirpation of whitebark pine.

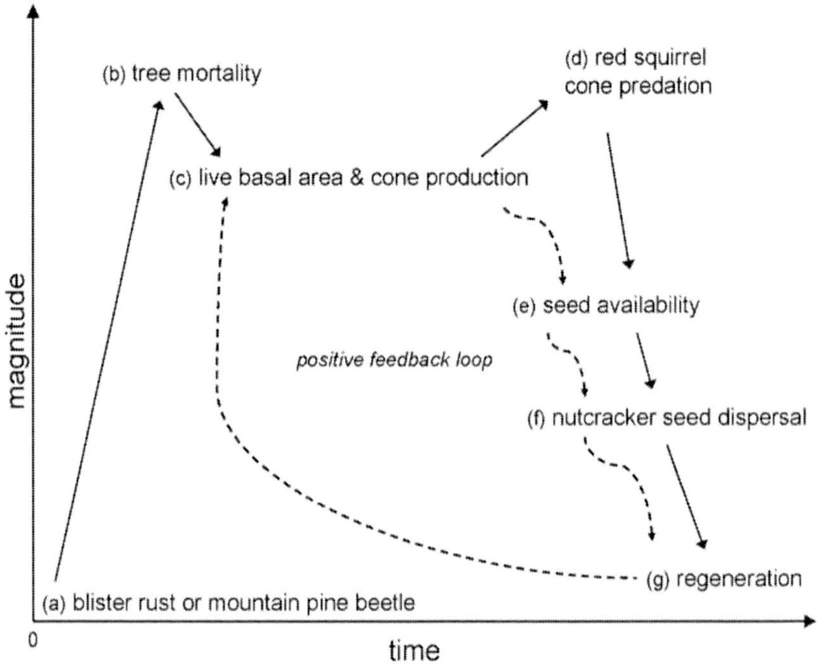

Figure 5. Analytically-based model depicting the series of processes between blister rust infection and or mountain pine beetle outbreak in a whitebark pine forest and the creation of a positive feedback loop hastening the decline of whitebark pine within the forest. Solid arrows preceding components indicate the direction of influence of the associated component. As tree mortality (b) increases, live basal area and cone production (c) decrease, and red squirrel cone predation (d) increases, leaving fewer seeds available (e) for nutcracker seed dispersal (f). Declining seed dispersal activity results in reduced whitebark pine regeneration (g). With fewer seedlings regenerating, basal area and cone production continue to decline over time, generating a positive feedback loop (dashed arrows). All components refer specifically to whitebark pine. Data are from 18 forest sites ranging in size from two to seven ha located in the Northern and Central Rocky Mountains, USA (2004-2006).

Implications for Whitebark Pine Management

In the Northern Rocky Mountains, where mortality is high and live whitebark pine abundance low (Figures 1 and 3), red squirrels can greatly diminish whitebark pine cone crops (Figure 4, Table 1). Our results also show a trajectory of decreasing interaction between nutcrackers and whitebark pine in the Northern Rocky Mountains, and point to decreased nutcracker seed-dispersal services as the principal mechanism behind lower levels of regeneration in the region (Table 1). Given the high levels of blister rust infection and tree mortality, and the low levels of live basal area documented, it appears that many whitebark pine forests in the NRM are no longer self-sustaining; making it apparent that active management will be needed to augment severely diminished potential for natural regeneration. Sites with < 5.0 m^2 ha^{-1} of live whitebark pine will require planting of rust-resistant seedlings, especially with further whitebark pine losses highly likely (McKinney et al. 2009).

The most comprehensive restoration treatments will be required in seral whitebark pine communities where cone predation, whitebark pine mortality, crown kill, rust infection, and advanced successional replacement of whitebark pine are the most dramatic. Declining whitebark pine communities are widespread in the NRM, resulting in part from past fire suppression activities in combination with decades-long presence of *Cronartium ribicola*. In these seral forests, restoration of whitebark pine will involve allowing some natural lightning-ignited fires to burn and applying silvicultural cuttings that remove encroaching shade-tolerant subalpine fir and Engelmann spruce—followed by planting blister rust-resistant seedlings (Keane and Arno 2001). These management actions may be the only way to maintain whitebark pine on the landscape. In forest stands that may equal or exceed the 5.0 m^2 ha^{-1} live whitebark pine basal area threshold, creating forest openings may encourage nutcracker seed dispersal and natural regeneration, which could spread genes from trees resistant to blister rust (Keane and Arno 2001, Schoettle 2004, Mahalovich et al. 2006).

Our data show that sites that exceed the 5.0 m^2 ha^{-1} threshold can still rely on nutcracker seed dispersal in some years, although these forests will lose whitebark pine over time as blister rust infection kills trees and damages canopies and as mountain pine beetle outbreaks continue. Managers are encouraged to identify such sites and use appropriate silvicultural treatments to increase nutcracker-caching habitat, at least for the immediate future. For example, research shows that blister rust has not yet caused as much damage in whitebark pine-dominant forests, which tend to be climax stands that are self-regenerating (Keane 2001, McKinney and Fiedler 2010). Given that squirrel predation is also low in these forests, and declines with increasing whitebark pine abundance (Figure 4), these forests can serve as seed sources for natural regeneration by nutcrackers and thus require less intensive management (e.g., allowing for wildland fire only). As an example, if a whitebark pine forest has > 5.0 m^2 ha^{-1} of live basal area, is at least 10 ha in area, and is not isolated from other whitebark pine forests, it could serve as a natural seed source for a restoration project. Removal (cutting) of competing shade-tolerant trees followed by prescribed burning at a location within 10 km of the whitebark pine seed source would likely attract nutcracker caching and increase the likelihood of natural regeneration (Keane and Arno 2001). Site-specific knowledge of whitebark pine forest attributes would also allow fire managers to make informed decisions regarding ''wildland fire use,'' which entails deciding when and where to allow lighting-ignited fires to burn. Wildland fires could be allowed to burn in subalpine forests where the probability of nutcracker seed dispersal is high and the potential for damage to humans and property is low.

Natural regeneration, facilitated by restoration treatments as described above, should result in a higher proportion of rust-resistant individuals relative to the parental generation because of ongoing mortality and damage from blister rust in susceptible trees, assuming that nutcracker seed dispersal occurs. In some remote locations, the use of wildland fire may be the best approach to whitebark pine restoration, given reasonably healthy seed sources. However, seedlings that are susceptible to blister rust have a poor chance of survival in areas of high blister rust infection levels, and the need to plant rust-resistant seedlings may be inevitable. Planting rust-resistant seedlings may well be the best strategy for spreading genetic resistance to blister rust as rapidly as possible and ensuring that whitebark pine will remain on the landscape.

Other complications need to be considered in whitebark pine management in the Central Rocky Mountains. Whitebark pine seeds are an important source of nutrition for grizzly bears in this region, where the bear has just been relisted by the U.S. Fish and Wildlife Service as a threatened species. Bears obtain seeds by raiding squirrel middens (Mattson and Reinhart 1997), and research shows that middens are significantly more likely to occur in mixed species forests containing whitebark pine (McKinney and Fiedler 2010). Our surveys in the CRM show that whitebark pine mortality is low (15%), but that current rust infection levels of nearly 50% portend future increases in mortality. Whitebark pine mortality at levels similar to those in the NRM

(70%) would have serious ramifications for the bear's status in the future. Because grizzly bears rely on red squirrels to access whitebark pine seeds, converting stands to whitebark pine dominant forest type, as discussed above for the NRM would not be a prudent option. Planting rust-resistant seedlings prior to severe decline and allowing high-elevation fires to burn when possible, maintaining mosaics of communities at different successional stages, are the most feasible options in the CRM.

ACKNOWLEDGMENTS

Funding was provided by a U.S. Forest Service Research Joint Venture Agreement, a U.S. National Park Service fellowship, a U.S. Department of Interior Cooperative Conservation Initiative grant, a U.S. Department of Agriculture McIntire-Stennis Research Program grant, and the College of Forestry and Conservation, University of Montana. We thank R. Keane, W. McCaughey, K. Tonnessen, and T. Carolin for logistical support. We are especially grateful to C. Fiedler for providing thoughtful comments on the manuscript and assistance throughout the study. We thank H. Zuuring, A. Sala, and E. Crone for sampling design and analysis consultations; and J. Fothergill, S. Sweeney, B. Cook, and B. Stauffer for field assistance.

REFERENCES

Arno, S.F. 2001. Community types and natural disturbance processes. In: *Whitebark Pine Communities: Ecology and Restoration.* Ed. by Tomback, D. F.; Arno, S. F.; Keane, R. E. Washington, D.C., USA: Island Press, pp. 74-88.

Arno, S. F., and Hammerly, R. P. 1984. *Timberline: Mountain and Arctic Forest Frontiers.* Seattle, Washington, USA: The Mountaineers. 304 p.

Arno, S.F., and Hoff, R. J. 1990. *Pinus albicaulis* Engelm. Whitebark pine. In: Silvics of North America, Volume 1, Conifers. Ed. by Burns, R.P.; Honkala, B.H. Washington, D.C., USA: *USDA Forest Service, Agriculture Handbook* 654, pp. 268-279.

Arno, S. F., and Allison-Bunnell, S. 2002. *Flames in our forest: disaster or renewal?* Washington, D.C., USA: Island Press, 227 pp.

Axelrod, D. I. 1986. Cenozoic history of some western American pines. *Annals Missouri Botanical Garden* 73:565-641.

Baud, K. S. 1993. *Simulating Clark's Nutcracker caching behavior: germination and predation of seed caches. Master's thesis*, University of Colorado, Denver, CO, USA.

Benkman, C. W., Balda R. P., and Smith, C. C. 1984. Adaptations for seed dispersal and the compromises due to seed predation in limber pine. *Ecology* 65:632-642.

Bock, W. J., Balda, R. P., and Vander Wall, S. B. 1973. Morphology of the sublingual pouch and tongue musculature in Clark's nutcracker. *Auk* 90:491-519.

Bruederle, L. P., Rogers, D. L., Krutovskii, K. V., and Politov, D. V. 2001. Population genetics and evolutionary implications. In: *Whitebark Pine Communities: Ecology and Restoration.* Ed. by Tomback, D. F., Arno, S. F., Keane, R. E. Washington, D.C., USA, Island Press, pp. 137-153.

Callaway, R. M. 1998. Competition and facilitation on elevation gradients in subalpine forests of the northern Rocky Mountains, USA. *Oikos* 2:561-573.

Campbell, E. M., and Antos, J. A. 2003. Postfire succession in *Pinus albicaulis – Abies lasiocarpa* forests of southern British Columbia. *Canadian Journal of Botany* 82:383-397.

Department of the Interior, Fish and Wildlife Service. 2010. Endangered and threatened wildlife and plants; 90-day finding on a petition to list *Pinus albicaulis* (whitebark pine) as endangered or threatened with critical habitat. *Federal Register* 75:42033-42040.

Elliott, P. F. 1974. Evolutionary responses of plants to seed-eaters: pine squirrel predation on lodgepole pine. *Evolution* 28:221-231.

Elliott, P. F. 1988. Foraging behavior of a central-place forager: field tests of theoretical predictions. *American Naturalist* 331:159-174.

Ellison, A. M., Bank, M. S., Clinton, B. D., Colburn, E. A., Elliott, K., Ford, C. R., Foster, D. R., Kloeppel, B. D., Knoepp, J. D., Lovett, G. M., Mohan, J., Orwig, D. A., Rodenhouse, N. L., Sobczak, W. V., Stinson, K. A., Stone, J. K., Swan, C. M., Thompson, J., Von Holle, B., and Webster, J. R. 2005. Loss of foundation species: consequences for the structure and dynamics of forested ecosystems. *Frontiers in Ecology and the Environment* 3:479-486.

Farnes, P. E. 1990. *SNOTEL and snow course data: Describing the hydrology of whitebark pine ecosystems.* In: Proceedings—Symposium on Whitebark Pine Ecosystems: Ecology and Management of a High-mountain

Resource. Compiled by Schmidt, W. C., McDonald, K. J. General Technical Report INT-270, Ogden, Utah, USA: USDA Forest Service, Intermountain Research Station, pp. 302-304.

Finley, R. B. Jr. 1969. *Cone caches and middens of Tamiasciurus in the Rocky Mountain region.* Miscellaneous Publications of the University of Kansas Museum of Natural History 51:233-273.

Flyger V., and Gates J. E. 1982. Pine squirrels: *Tamiasciurus hudsonicus, T. douglasii.* In: *Wild mammals of North America.* Ed. by Chapman J. A., Feldhamer G. A. John Hopkins University Press, Baltimore, Maryland, pp. 230-237.

Furnier, G. R., Knowles, P., Clyde, M. A., and Dancik, B. P. 1987. Effects of avian seed dispersal on the genetic structure of whitebark pine populations. *Evolution* 41:607-612.

Geils, B. W., Hummer, K. E., and Hunt, R. S. 2010. White pines, *Ribes*, and blister rust, a review and synthesis. *Forest Pathology* 40:147-185.

Gernandt, D. S., Geada López, G. G., Ortiz Garcia, S., and Liston, A. 2005. Phylogeny and classification of *Pinus*. *Taxon* 54:29-42.

Gibson, K., Skov, K., Kegley, S., Jorgensen, C., Smith, S., and Witcosky, J. 2008. *Mountain pine beetle impacts in high-elevation five-needle pines: Current trends and challenges.* Forest Health Protection R1-08-020. Missoula, MT, USA: USDA, Forest Service.

Gilpin, M. E., and Soulé, M. E. 1986. Minimum viable populations: processes of species extinction. In: *Conservation Biology: the science of scarcity and diversity*. Ed. by Soulé, M. E. Sinauer Associates, Inc., Sunderland, Massachusetts.

Gurnell, J. 1984. Home range, territoriality, caching behaviour and food supply of the red squirrel (*Tamiasciurus hudsonicus fremonti*) in a subalpine lodgepole pine forest. *Animal Behavior* 32:1119-1131.

Hafner, D. J. 1984. Evolutionary relationships of the nearctic Sciuridae. In: *The biology of ground dwelling squirrels*. Ed. by Murie J. O., Michener G. R. University of Nebraska Press, Lincoln, Nebraska, pp. 3-23.

Hamann, A., and Wang, T. 2006. Potential effects of climate change on ecosystem and tree species distribution in British Columbia. *Ecology* 87:2773-2786.

Hutchins, H. E., and Lanner, R. M. 1982. The central role of Clark's nutcracker in the dispersal and establishment of whitebark pine. *Oecologia* 55:192-210.

Kamil, A. C., and Balda, R. P. 1985. Cache recovery and spatial memory in Clark's nutcracker (*Nucifraga columbiana*). *Journal of Experimental Psychology: Animal Behavior Processes* 11:95-111.

Keane, R. E. 2001. Successional dynamics: modeling an anthropogenic threat. In: *Whitebark pine communities: Ecology and restoration.* Ed. by Tomback D. F., Arno S. F., Keane R. E. Island Press, Washington, D.C., pp. 159-192.

Keane, R. E., and Arno, S. F. 2001. Restoration concepts and techniques. In: *Whitebark pine communities: Ecology and restoration.* Ed. by Tomback D. F., Arno S. F., Keane R. E. Island Press, Washington, D.C., pp. 367-400.

Keane, R. E., Ryan, K. C., Veblen, T. T., Allen, C. D., Logan, J. A., and Hawkes, B. 2002. Cascading effects of fire exclusion in Rocky Mountain ecosystems. In*: Rocky Mountain Futures: An Ecological Perspective.* Ed. by Baron, J. S. Washington, D.C., USA: Island Press, pp. 133-152.

Kendall, K. C., and Keane, R. E. 2001. Whitebark pine decline: infection, mortality, and population trends. In: *Whitebark Pine Communities: Ecology and Restoration.* Ed. by Tomback, D. F., Arno, S. F., Keane, R. E. Washington, D.C., USA, Island Press, pp. 221-242.

Lanner, R. M. 1982. Adaptations of whitebark pine for seed dispersal by Clark's nutcracker. *Canadian Journal of Forest Research* 12:391-402.

Lanner, R. M. 1990. *Biology, taxonomy, evolution, and geography of stone pines of the world.* In: Proceedings–Symposium on Whitebark Pine Ecosystems. Compiled by Schmidt, W. C., McDonald, K. J. General Technical Report INT-270. Ogden, Utah, USA: USDA Forest Service, Intermountain Research Station, pp. 14-24.

Lanner, R. M., and Gilbert, B. K. 1994. *Nutritive value of whitebark pine seeds, and the question of their variable dormancy.* In: Proceedings— International Workshop on Subalpine Stone Pines and Their Environment: the Status of Our Knowledge. Compiled by Schmidt, W. C., Holtmeier, F. K. General Technical Report INT-GTR-309, Ogden, Utah, USA. USDA Forest Service, Intermountain Research Station, pp. 206-211.

Larsen K. W., and Boutin S. 1994. Movements, survival, and settlement of red squirrel (*Tamiasciurus hudsonicus*) offspring. *Ecology* 75:241-223.

Linhart, Y. B., and Tomback, D. F. 1985. Seed dispersal by Clark's nutcracker causes multi-trunk growth form in pines. *Oecologia* 67:107-110.

Liston, A., Robinson, W. A., Piñero, D., and Alvarez-Buylla, E. R. 1999. Phylogenetics of *Pinus* (Pinaceae) based on nuclear ribosomal DNA internal transcribed spacer region sequences. *Molecular Phylogenetics and Evolution* 11:95-109.

Logan, J. A., and Powell, J. A. 2001. Ghost forests, global warming, and the mountain pine beetle (Coleoptera: Scolytidae). *American Entomologist* 47:160-172.

Logan, J. A., Régnière, J., and Powell, J. A. 2003. Assessing the impacts of global warming on forest pest dynamics. *Frontiers in Ecology and the Environment* 1:130-137.

Logan, J. A., MacFarlane, W. W., and Willcox, L. 2010. Whitebark pine vulnerability to climate-driven mountain pine beetle disturbance in the Greater Yellowstone Ecosystem. *Ecological Applications* 20: 895-902.

Lorenz, T. J., and Sullivan, K. A. 2009. Seasonal differences in space use by Clark's Nutcrackers in the Cascade Range. *Condor* 11:326-340.

Mahalovich M. F., Burr, K. E., and Foushee, D. L. 2006. *Whitebark pine germination, rust resistance, and cold hardiness among seed sources in the Inland* Northwest: Planting strategies for restoration. Forest Service Proceedings RMRS-P-43.

Mattes, H. 1978. Der Tannenhäher (*Nucifraga caryocatactes* L) im engadin: Studien zu seiner Ökologie und Function im Arvenwald. *Münsterische Geographische Arbeiten* 2.

Mattson D. J., and Reinhart, D. P. 1997. Excavation of red squirrel middens by grizzly bears in the whitebark pine zone. *Journal of Applied Ecology* 34:926-940.

McCaughey, W. W. 1990. Biotic and microsite factors affecting *Pinus albicaulis* establishment and survival. *Doctoral Dissertation*, Montana State University, Bozeman, MT.

McCaughey, W. W., and Schmidt, W. C. 1990. *Autecology of whitebark pine*. In: Proceedings—Symposium on Whitebark Pine Ecosystems: Ecology and Management of a High-mountain Resource. Compiled by Schmidt, W. C., McDonald, K. J. General Technical Report INT-270, Ogden, Utah, USA: USDA Forest Service, Intermountain Research Station, pp. 85-96.

McCaughey, W. W., and Schmidt, W. C. 2001. Taxonomy, distribution, and history. In: *Whitebark Pine Communities: Ecology and Restoration*. Ed. by Tomback, D. F.; Arno, S. F.; Keane, R. E. Washington, D.C., USA: Island Press, pp. 29-40.

McCaughey, W. W., and Tomback, D. F. 2001. The natural regeneration process. In: *Whitebark Pine Communities: Ecology and Restoration*. Ed. by Tomback, D. F., Arno, S. F., Keane, R. E. Washington, D.C., USA: Island Press, pp. 105-120.

McDonald, G. I., and Hoff, R. J. 2001. Blister rust: an introduced plague. In: *Whitebark Pine Communities: Ecology and Restoration.* Ed. by Tomback, D. F., Arno, S. F., Keane, R. E. Washington, D.C., USA, Island Press, pp. 193-220.

McKenny, D. W., Pedlar, J. H., Lawrence, K., Campbell, K., and Hutchinson, M. F. 2007. Potential impacts of climate change on the distribution of North American trees. BioScience 57:939-948. Supplementary material accessed from *http://planthardiness.gc.ca/* 5 July 2008.

McKinney, S. T., and Tomback, D. F. 2007. The influence of white pine blister rust on seed dispersal in whitebark pine. *Canadian Journal of Forest Research* 37:1044-1057.

McKinney, S. T., Fiedler, C. E, and Tomback, D. F. 2009. Invasive pathogen threatens bird-pine mutualism: implications for sustaining a high-elevation ecosystem. *Ecological Applications* 19:597-607.

McKinney, S. T., and Fiedler, C. E. 2010. Tree squirrel habitat selection and predispersal seed predation in a declining subalpine conifer. *Oecologia* 162:697-707.

Mills, L. S., Soulé, M. E., and Doak, D. F. 1993. The keystone-species concept in ecology and conservation. *BioScience* 43:219-224.

Morin, X., and Thuiller, W. 2009. Comparing niche- and process-based models to reduce prediction uncertainty in species range shifts under climate change. *Ecology* 90:1301-1313.

Ogilvie, R. T. 1990. *Distribution and ecology of whitebark pine in western Canada. In: Proceedings–Symposium on Whitebark Pine Ecosystems.* Compiled by Schmidt, W. C., McDonald, K. J. General Technical Report INT-270. Ogden, Utah, USA: USDA Forest Service, Intermountain Research Station, pp. 54-60.

Perkins, D. L., and Swetnam, T. W. 1996. A dendroecological assessment of whitebark pine in the Sawtooth-Salmon River region Idaho. *Canadian Journal of Forest Research* 26:2123-2133.

Price, R. A., Liston, A., and Strauss, S. H. 1998. Phylogeny and systematics of *Pinus*. In: *Ecology and Biogeography of Pinus.* Ed. by Richardson, D. M. Cambridge, UK. Cambridge University Press, pp. 49-68.

Resler, L. M. 2004. *Conifer establishment sites on a periglacial landscape*, Glacier National Park, Montana. Ph.D. Dissertation, Department of Geography, Texas State University-San Marcos.

Resler, L. M., and Tomback, D. F. 2008. Blister rust prevalence in krummholz whitebark pine*: Implications for treeline dynamics. Arctic, Antarctic, and Alpine Research* 40:161-170.

Richardson, B. A., Brunfeld, S. J., and Klopfenstein, N. B. 2002. DNA from bird-dispersed seed and wind-disseminated pollen provides insights into postglacial colonization and population genetic structure of whitebark pine (*Pinus albicaulis*). *Molecular Ecology* 11:215-227.

Rogers, D. L., Millar, C. I., and Westfall, R. D. 1999. Fine-scale genetic architecture of whitebark pine (*Pinus albicaulis*): Associations with watershed and growth form. *Evolution* 53:74-90.

Rusch, D. A., and Reeder W. G. 1978. Population ecology of Alberta red squirrels. *Ecology* 59:400-420.

Schoettle, A. W. 2004. Developing proactive management options to sustain bristlecone and limber pine ecosystems in the presence of a non-native pathogen. In: *Silviculture in special places:* Proceeding of the National Silviculture Workshop. Compiled by Shepperd, W. D., Eskew, L. G. Proceedings RMRS-P-34, U.S. Dept Agric, Rocky Mountain Research Station, Fort Collins, Colorado, pp. 146-155.

Schrag, A. M., Bunn, A. G., and Graumlich, L. J. 2008. Influence of bioclimatic variables on treeline conifer distribution in the Greater Yellowstone Ecosystem: implications for species of conservation concern. *Journal of Biogeography* 35:698-710.

Schwandt, J. W., Lockman, I. B., Kliejunas, J. T., and Muir, J. A. 2010. Current health issues and management strategies for white pines in the western United States and Canada. *Forest Pathology* 40:226-250.

Sibley, C. G., and Ahlquist, J. E. 1985. The phylogeny and classification of the Australo-Papaun passerine birds. *Emu* 85:1-14.

Siepielski, A. M., and Benkman, C. W. 2007a. Convergent patterns in the selection mosaic for two North American bird-dispersed pines. *Ecological Monographs* 77:203-220.

Siepielski, A. M., and Benkman, C. W. 2007b. Selection by a predispersal seed predator constrains the evolution of avian seed dispersal in pines. Functional Ecology 21:611-618.

Smith, C. C. 1970. The coevolution of pine squirrels (*Tamiasciurus*) and conifers. *Ecological Monographs* 40:349-371.

Smith, C. C. 1981. The indivisible niche of *Tamiasciurus*: an example of nonpartitioning of resources. *Ecological Monographs* 51:343-363.

Smith, C. M., Wilson, B., Rasheed, S., Walker, R. C., Carolin, T., and Sheppard, B. 2008. Whitebark pine and white pine blister rust in the Rocky Mountains of Canada and northern Montana. *Canadian Journal of Forest Research* 38:982-995.

Soulé, M. E., Estes, J. A., Berger, J., and Martinez del Rio, C. 2003. Ecological effectiveness: conservation goals for interactive species. *Conservation Biology* 17:1238-1250.

Steele M. A., Wauters L. A., and Larsen, K. W. 2005. Selection, predation, and dispersal of seeds by tree squirrels in temperate and boreal forests: are tree squirrels keystone granivores? In: *Seed fate: predation, dispersal and seedling establishment*. Ed. by Forget P. M., Lambert, J. E., Hulme P. E., Vander Wall, S. B. CABI, Wallingford, Oxon, UK pp. 205-221.

Taylor, S.W., and Carroll, A. L. 2004. Disturbance, forest age, and mountain pine beetle outbreak dynamics in BC: a historical perspective. In: *Challenges and Solutions. Proceedings of the Mountain Pine Beetle Symposium*. Ed. by Shore, T. L., Brooks, J. E., Stone, J. E. Information Report BC-X-399, Victoria, B.C., Canada: Natural Resources Canada, Canadian Forest Service, Pacific Forestry Centre. pp. 41-56.

Tillman-Sutela, E., Kauppi, A., Karppinen K., and Tomback, D. F. 2008. Variant maturity in seed structures of *Pinus albicaulis* (Engelm.) and *Pinus sibirica* (Du Tour): key to a soil seed bank, unusual among conifers? *Trees* 22:225-236.

Tomback, D. F. 1978. Foraging strategies of Clark's nutcracker. *Living Bird* 16:123-161.

Tomback, D. F. 1982. Dispersal of whitebark pine seeds by Clark's nutcracker: A mutualism hypothesis. *Journal of Animal Ecology* 51:451-467.

Tomback, D. F. 1986. Post-fire regeneration of krummholz whitebark pine: a consequence of nutcracker seed caching. *Madroño* 33:100-110.

Tomback, D. F. 1988. *Nutcracker-pine mutualisms: multi-trunk trees and seed size*. Acta XIX Congressus Internationalis Ornithlogici, vol 1. Ed. by Ouellet, H. University of Ottawa Press, Ottawa, Ontario, Canada, pp. 518-527.

Tomback, D. F. 1998. Clark's Nutcracker (*Nucifraga columbiana*). In: *The birds of North America*. Ed. by Poole, A., Gill F. The Birds of North America Incorporated, Philadelphia, Pennsylvania, Number 331.

Tomback, D. F. 2001. Clark's nutcracker: agent of regeneration. In: *Whitebark Pine Communities: Ecology and Restoration*. Ed. by Tomback, D. F., Arno, S. F., Keane, R. E. Washington, D.C., USA: Island Press, pp. 89-104.

Tomback, D. F. 2005. The impact of seed dispersal by Clark's nutcracker on whitebark pine: multi-scale perspective on a high mountain mutualism. In:

Mountain Ecosystems: Studies in Treeline Ecology. Ed. by Broll, G., Keplin, B. Berlin, Germany: Springer, pp. 181-201.

Tomback, D. F., and Linhart, Y. B. 1990. The evolution of bird-dispersed pines. *Evolutionary Ecology* 4:185-219.

Tomback, D. F., and Kendall, K. C. 2001. Biodiversity losses: the downward spiral. In: *Whitebark Pine Communities: Ecology and Restoration*. Ed. by Tomback, D. F., Arno, S. F., Keane, R. E. Washington, D.C., USA: Island Press, pp. 243-262.

Tomback, D. F., and Achuff, P. 2010. Blister rust and western forest biodiversity: ecology, values and outlook for white pines. *Forest Pathology* 40:186-225.

Tomback, D. F., Hoffmann, L. A., and Sund, S. K. 1990. *Coevolution of whitebark pine and nutcrackers: implications for forest regeneration*. In: Proceedings—Symposium on Whitebark Pine Ecosystems: Ecology and Management of a High-mountain Resource. Compiled. By Schmidt, W. C., McDonald, K. J. General Technical Report INT-270, Ogden, Utah, USA: USDA Forest Service, Intermountain Research Station, pp. 118-129.

Tomback, D. F., Sund, S. K., and Hoffmann, L. A. 1993. Post-fire regeneration of *Pinus albicaulis*: Height-age relationships, age structure, and microsite characteristics. *Canadian Journal of Forest Research* 23:113-119.

Tomback, D. F., Arno, S. F., and Keane, R. E. 2001a. The compelling case for management intervention. In: *Whitebark Pine Communities: Ecology and Restoration*. Ed. by Tomback, D. F., Arno, S. F., Keane, R. E. Washington, D.C., USA: Island Press, pp. 3-25.

Tomback, D. F., Anderies, A. J., Carsey, K. S., Powell, M. L., and Mellmann-Brown, S. 2001b. Delayed seed germination in whitebark pine and regeneration patterns following the Yellowstone fires. *Ecology* 82:2587-2600.

Vander Wall, S. B. 1982. An experimental analysis of cache recovery in Clark's Nutcracker. *Animal Behavior* 30:84-94.

Vander Wall, S. B., and Balda, R. P. 1977. Coadaptations of the Clark's nutcracker and piñon pine for efficient seed harvest and dispersal. *Ecological Monographs* 47:89-111.

Vander Wall, S. B, and Hutchins, H. E. 1983. Dependence of Clark's Nutcracker, *Nucifraga columbiana*, on conifer seeds during postfledging period. *Canadian Field-Naturalist* 97:208-214.

Warwell, M. V., Rehfeldt, G. E., and Crookston, N. L. 2007. *Modeling contemporary climate* In: Proceedings of the conference whitebark pine: a Pacific Coast perspective 2006. Coordination by Goheen, E. M., Sniezko, R. A. August 27-31, Ashland, OR, R6-NR-FHP-2007-01. Portland, OR USA: Pacific Northwest Region, Forest Service, U.S. Department of Agriculture, pp. 139-142.

Weaver, T. 2001. Whitebark pine and its environment. In: *Whitebark Pine Communities: Ecology and Restoration*. Ed. by Tomback, D. F., Arno, S. F., Keane, R. E. Washington, D.C., USA: Island Press, pp. 41-73.

Reviewed by Carl Fiedler, College of Forestry and Conservation, University of Montana.

In: Mountain Ecosystems
Editor: Kevin E. Richards, pp. 79-92
ISBN 978-1-61209-306-2

Chapter 3

MOUNTAINS ECOSYSTEMS AS A TEMPORAL SINK FOR PERSISTENT ORGANIC POLLUTANTS

***Ricardo Barra*[1] *and Roberto Quiroz*[2]**
[1]EULA-Chile Environmental Sciences Centre,
University of Concepción, Concepción; Chile
[2]Patagonian Ecosystems Research Centre CIEP, Coyhaique; Chile

ABSTRACT

In this chapter the role of mountain ecosystems in the dynamics and fate of Persistent Organic Pollutants (POPs) will be described. POPs are a group of chemicals released into the environment by anthropogenic processes, characterized by an elevated persistence, high bioaccumulation and toxicity. Recent research has shown an increasing trend of POP levels in mountain ecosystems all over the world, in various environmental media such as soils, ice, snow, vegetation and animals. Short term POP accumulation in snow and ice may occur in different mountain systems depending on atmospheric circulation, temperature patterns and precipitation rates. On a long term basis accumulation of POPs in mountain lake sediments may be used for reflecting temporal and spatial trends in deposition. Passive air sampling and bioindicators are now being used for demonstrating altitudinal trends, where cold condensation processes may cause higher levels of some POPs at higher altitudes. Geographical trends will be also discussed, considering research performed in the major mountain ranges around the world (the Alps, Pyrinees, Andes, Rocky Mountains etc.).

INTRODUCTION

The general population perceive that the polar and mountain areas are pristine places, free of contamination, a last frontier where human activity has not had much effect. However, in the last two decades it have been shown that the Antarctic, the Arctic and high mountains (Andes, Alps, Pyrenees) are generally accumulating certain contaminants, some of them of exclusively human origin, which are known generically as Persistent Organic Pollutants (POPs). These are produced and released by the human activity, and are characterized by: elevated persistence (low degradation rates), ability to accumulate in the fatty tissue of organisms, toxic effects at the neurological, endocrine and immune systems, and the potential to travel long distances by atmospheric transport or by other vectors (including birds and large mammals)

Among the POPs are compounds that have not been manufactured since the seventies, such as some organochlorine pesticides (aldrin, heptachlor), industrial products (polychlorinated biphenyls (PCBs) and hexachlorobenzene (HCB)) and non intentional release chemicals (such as dioxins and furans). The Stockholm Convention on POPs was internationally endorsed in 2001 (www.pops.int). Through this agreement, respective countries agreed to eliminate those pollutants of synthetic origin, and reduce the release of those contaminants that are produced unintentionally such as dioxins and furans, and also take measures to prevent the continued environmental contamination with these compounds.

Mountains Environments serve as early warning systems both for climate change and pollution by persistent organic pollutants, in particular because mountains are located near the atmospheric sources, and could therefore serve as monitors of the levels of POPs in the environment

PREVIOUS REPORTS ON POPS IN MOUNTAIN AREAS

Similar to Arctic environments, high-altitude mountainous ecosystems have been considered as the last pristine and ''untouched'' environments on our globe. However, recent studies have revealed that alpine environments, similar to the polar regions, are also affected by the transport of POPs (Blais et al., 1998; Sarkar et al., 2003). The first evidence for the presence of organic contaminants in elevated ecosystems dates back to the 1970s (Reid,1978). However, no comprehensive follow-up study was performed until the 1990s, when simultaneous North American and European transport and fate studies of POPs were performed in high-altitude environments (Calamari et al., 1991; Camarero et al., 1995; Blais et al., 1998; Datta et al., 1998; Carrera et al., 2001; Vilanova et al., 2001; Quiroz et al., 2008) and more recently in the Andes (Barra et al., 2005; Borghini et al., 2005; Pozo et al., 2007; Quiroz et al., 2009) and Himalayas (Wang et al., 2006; Kang et al., 2009; Wang et al., 2010).

Canadian scientists first pinpointed the implication of POPs for high-altitude environments in a comprehensive way in the late 1990s, (Blais et al.,1998). Under certain meteorological and geographic conditions, high-altitude environments can serve as ‘‘cold condensers’’ for atmospheric POP loadings. Subsequent investigations in high-altitude environments in Asia, Europe, and North and South America have confirmed suspicions that high-altitude mountainous regions have the potential to serve as focus regions for POPs and even for non persistent, moderately persistent contaminants, such as currently used pesticides, due to cold condensation and deposition in high altitudes. Although the presence and the altitude-dependent increase of POP levels in mountainous regions are confirmed by many international studies, the ecotoxicological consequences still remain largely unknown.

POP Cycling in Mountain Zones

In general, mountain areas are characterized by low temperatures and high rainfall compared to low altitudes. This would favor the transport and accumulation of POPs from the valleys to higher elevations. The direction of atmospheric transport is determined by the movement pattern of air masses, followingthe global atmospheric circulation pattern, modified by the presence of mountains. Figure 1b shows a simplified diagram of the movement of air masses and the effect of mountains. In general, atmospheric circulation is determined by pressure difference of air masses and the Coriolis effect (caused by the rotation of the Earth,Figure 1 a) giving rise to a movement from low to high latitudes. This scheme is in reality much more complex due to other factors such as the influence of ocean currents, the center of high and low pressure, the presence of mountains, etc.

As a complement to the air masses global circulation, local and regional processes take place that favor the accumulation of POPs in mountain areas.In general the movement of air masses is caused by temperature differences between the high and lowland areas or between mountains. As shown in Figure 2, during the day the temperature gradient allows the mobilization of the air masses from the valleys to the high altitudes (Figure 2 a), whereas during the night, the temperature gradient decreases causing a reverse in direction (Figure 2 b). This system may have a range of influence of several hundred kilometers (Prévôt et al., 2000).

POPs in the atmosphere, either in gaseous form or associated with the particulate matter, therefore are displaced by the movement of air masses into the high mountain areas during the day and reverse direction at night. However, this cycle is compounded by the effect of temperature and precipitation, and the net balance is the retention of POPs in different environmental compartments of the high mountain areas likely to accumulate these contaminants (Figure 2). On the one hand, high temperatures typical of low altitudes favor volatilization processes of these semi-volatile compounds from soil, vegetation or water into the atmosphere and therefore, making them available for transport to the high mountain areas. In turn, lower temperatures and increased frequency and intensity of precipitation in mountains, favors the process of retention of POPs from the atmosphere viadeposition and condensation (Figure 2), and in that way cannot be transported back to the valleys during the night conditions.

Recently, Daly and Wania (2005) have stated that the study of POPs in mountain areas is warranted for three basic reasons:

a) The assessment of the impact on the population: since mountains are a source of water supply for the population, pollution of snow or water from lakes can affect water quality, both for consumption and agricultural use.
b) The assessment of the impact on high mountain ecosystem: since these areas are characterized by communities of plants and animals of great ecological value.
c) The study of POPs in mountain areas would help improve knowledge about the POPs dynamics in the environment, which would help identify and assess factors affecting their fate and transport. An understanding of the influence of environmental factors (climate,

vegetation, temperature, precipitation) affecting the altitudinal distribution of POPs in mountain areas can help elucidate the mechanisms that affect large-scale distribution.

Globally, mountain areas represent approximately 27% of land area and 22% of the world's population (UNEP, 2002). Moreover, unlike the polar regions, mountains are often near densely populated and / or industrialized zones. For example, the U.S. Rocky Mountains are close to areas with important agricultural activity; the European Alps divide highly industrialized areas in southern Germany, Switzerland, Austria and Northern Italy; the Pyrinees are located near industrial areas of Spain and France; the Hymalayas are adjacent to densely populated areas, which is also true for the highlands in the Andes in South America. Through studies on pollutants in mountain environments, we promote an approach that allows the study of transport and dynamics of a greater number of families of pollutants compared with the global distribution studies focused on polar areas (Wania et al., 1999; Daly and Wania, 2005).

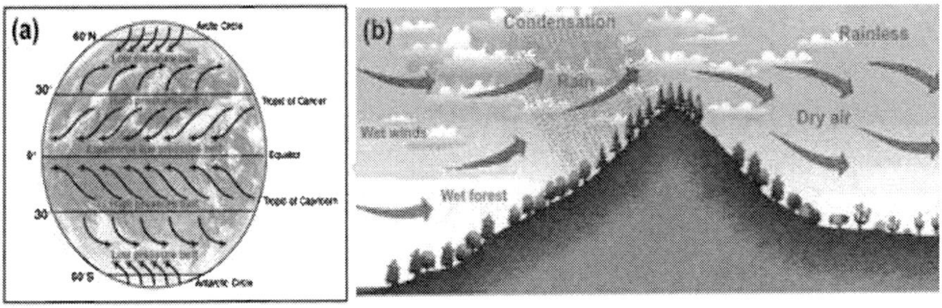

Figure 1. General scheme of globaslair mass movement (a); effect of mountain air masses(b).

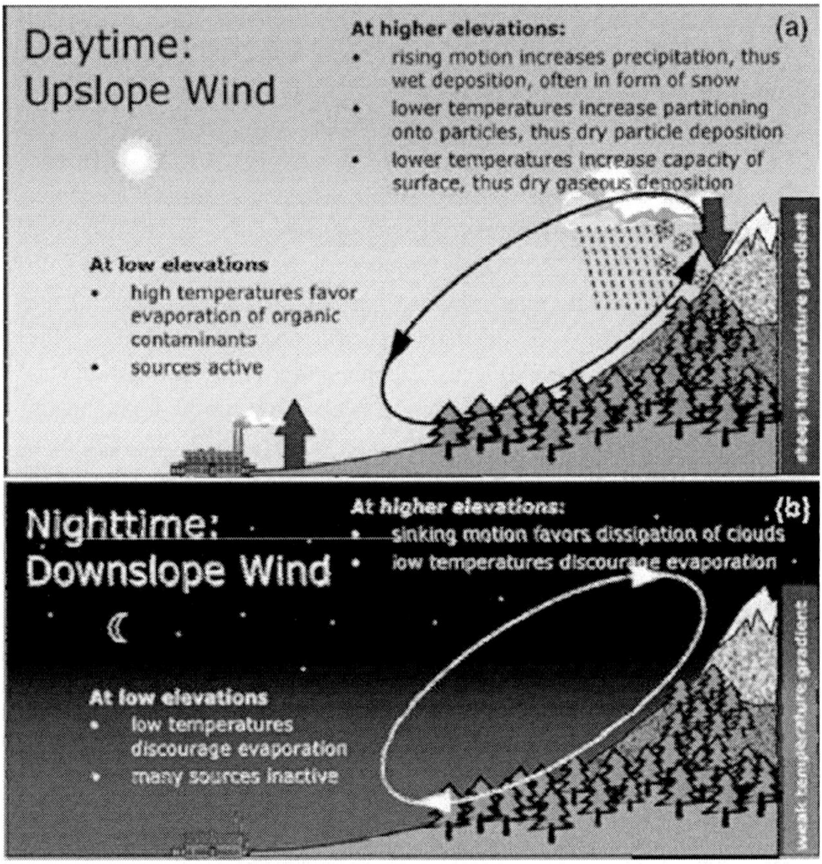

Based on Daly and Wania, (2005).

Figure 2. Influence of meteorological variables in the transport and accumulation of POPs in high mountain areas.

AIR

Atmospheric monitoring of POPs traditionally relies on high volume air samplers. However, active air sampling requires trained operators and a supply of electricity which is rarely found at truly remote mountain sites. In recent years, various types of passive air samplers (PAS) have been developed and used for POPs (Shoeib and Harner, 2002). Passive air samplers (PAS) consist of an accumulating medium that has a high retention capacity for the target analytes. Types of PAS most frequently used include polyurethane foams

(PUF), styrene divinyl benzene (XAD) and Semipermeable membrane devices (SPMD) types, where the main operational difference is that the exposure time should be sufficient to reach equilibrium state (Figure 3). This strategy has been used in different global- scale and regional studies where sites have been included in mountain areas, and permits the spatial representation of POPs in the environment.

The Global Atmospheric Passive Sampling (GAPS) study aims to demonstrate the feasibility of using passive samplers to assess the spatial distribution of persistent organic pollutants on areas on a worldwide basis. The GAPS network includes more than 40 sites on 7 continents, mainly in background locations, with some representation of urban and agricultural and mountains areas (Pozo et al., 2006; Pozo et al., 2008). In this study mountain sites were concentrated in Chilean and Bolivian Andes with altitudes > 4000 masl. In all of these sites more than five pesticides were detected in some cases in very high concentrations, as was the case for the organochlorine insecticide endosulfan.

Another study is from the Tibetan plateau, which has an area of over 2.5×10^6 km^2 and an average elevation of >4000 m, therefore it is known as "the third pole of the world" and is the highest and most extensive plateau on Earth. The atmosphere over the Tibetan Plateau remains relatively pristine due to sparse human population and minimal industrial activities. Air masses over the Tibetan Plateau are mainly dominated by continental air from central Asia and maritime air from the Indian Ocean. XAD 2-resin based passive air samplers were therefore deployed for 1 year (between July 2007-June 2008) at 16 locations.

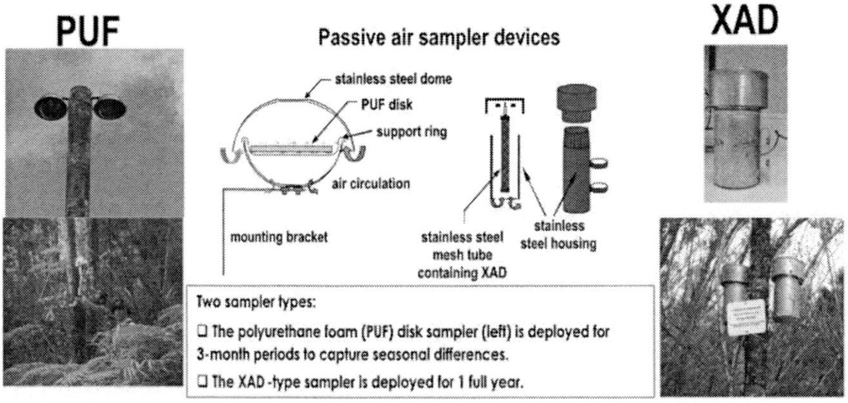

Figure 3. Schematic diagrams and photography of passive air samplers.

The results for air concentrations (pg/m^3) ranged as follows: DDTs, 5-75; HCHs, 0.1-36; R-endosulfan, 0.1-10; HCB, 2.8-80; sum of 15 PCBs, 1.8-8.2; and sum of 9 PBDEs, 0.1-8.3. The highest DDT concentrations occurred at Qamdo, where the sampling site is near farm land, indicating the spatial distribution of DDTs across the plateau may be influenced by scattered local usage of DDT. Higher levels of HCHs were observed at sites with high elevation (>4000 m) and close to the China-India border, indicating possible long-range atmospheric transport (Wang et al., 2010).

Snow

Because of their large and cold surface area, snowflakes are very effective at scavenging semi-volatile organic pollutants from the air and depositing them on the ground. Once on the ground, some of these contaminants may volatilize back to the air as the snowpack matures, while some might remain for later scavenging by organic detritus, or become dissolved in melt water and return to the soil as the snow melts.

Snowpack and glacial horizons are heterogeneous media, and considerable exchange of POPs takes place between snow and air during snowpack maturation and firnification (conversion to ice). Photolytic transformation in snow can dramatically affect chemical composition of pollutants. Vertical distributions of contaminants in snowpack reflect variations in deposition, and are significantly modified by post-depositional processes (Wania et al., 1998; Wania et al., 1999).

Glacial ice may provide a wealth of spatial and temporal information on POPs, with the principaladvantages of glacial ice depositionhaving high temporal resolution. With annual accumulation of ass much as several of meters, we can use glacial ice to resolve short-term seasonal patterns. Glacial profiles can be dated using well-established methods such as annual dark layers, tritium, volcanic tephras, Saharan dust, and stable isotopes.

There are still some knowledge gaps for understanding the fate of POPs in snowpacks (Figure 4), such as the role of wind, but it is expected that more research will be conducted in this area in the next years. Theoretical (Wania, 1997; Daly and Wania, 2004) and field studies (Herbert et al., 2005) suggest that the main channels through which POPs are initially lost from the mass of snow, is in the reduction of the specific area of snow and the decreased sorption coefficient on the surface of the snow due to increased temperature.

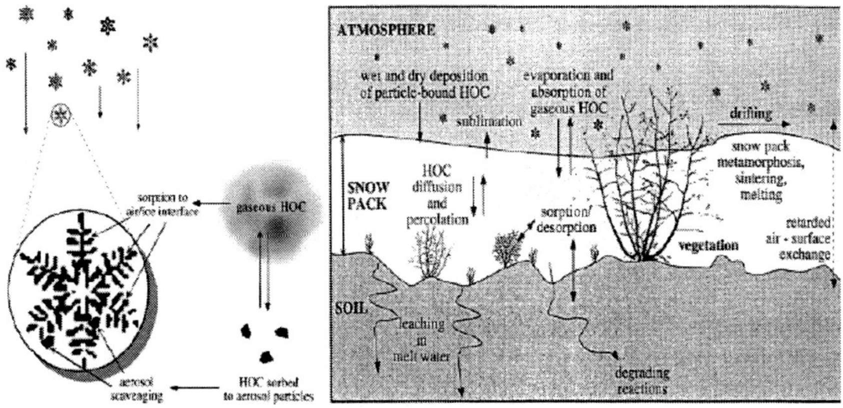

Taken from Wania et al., (1998).

Figure 4. Processes involved in the dynamics and fate of POPs in snow.

Atmospheric POPs trapped in snow, either by binding to aerosols and drawn into the snow, by adsorption to the crystal surface or by dissolution in droplets and freezing (Figure 4), subsequently accumulate as snow pack is deposited. The POP content changes owing to the multiple processes involved in ripening, such as sublimation or exchange between snow and atmosphere, or the longitudinal diffusion or depth or leakage to the ground (Figure 4) . During these processes, there is a reduction of the burden of POPs in relation to fresh snow and the loss of more volatile POPs due to evaporation from the gas phase to the atmosphere (Herbert et al., 2005).

Lake Sediments

Lake sediments have been very useful for determining the role of temperature and other environmental factors in the atmospheric deposition of POPs in mountain ecosystems. The behavior of POPs is governed by their physico-chemical properties but also the characteristics of the environment where they are released. In mountain lake sediments the law of superposition may apply (Smol, 2002) which means that for undisturbed sediments the deepest deposits are the oldest while the recent sediments are superficial, resulting in a depth time profile. If a pollutant is deposited in the sediments as described in Figure 5, a time dependent deposition may result. This is a key to understanding the fate and temporal depositional trends of POPs in mountain areas, and has been widely used for reconstructing historical pollution events in mountain ecosystems.

Remote high elevation lakes also seem to be more vulnerable to this kind of pollution because climatic conditions (cool environments) may enhance the residence time of POPs in these abiotic regions (Grimalt et al., 2001). Lake sediments are therefore widely used to reconstruct temporal profiles of environmental contamination (Pozo et al., 2007). Undisturbed sediment cores thus provide insight into local and global trends of past and present use of a wide range of substances, including the organochlorine compounds (Muir et al., 1995; Barra et al., 2006) For instance, studies examining sediments collected from remote lakes in the Canadian Arctic have detected organochlorine pesticides and PCBs (Muir et al., 1995; Muir et al., 1996).

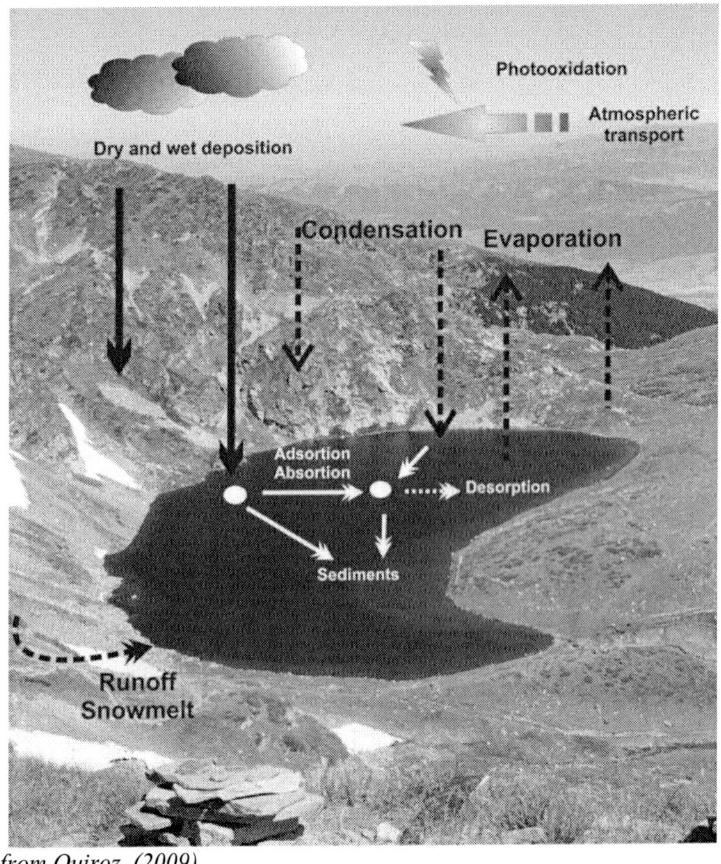

Taken from Quiroz, (2009).

Figure 5. *Processes* that govern the *fate* of *POPs* in mountains lacustrine *systems.*

Conclusions

Contrary to the common belief of mountain ecosystems as pristine areas, mountains are particularly affected by anthropogenic pollution in by synthetic chemicals such as POPs and related compounds. This is partially explained by the location of large urban centers near mountain areas, which act as emission sources, and by the natural characteristics of the mountain areas such as winds and temperature, resulting in elevated levels of persistent organic pollutant deposits in such areas.

Mountains are suitable locations for monitoring to assess both regional and global contamination of POPs. In fact many relevant mountain systems are located in temperate areas close to highly populated zones, therefore the potential of pollution not only for POPs but other less persistent contaminants, such as currently used pesticides may be of importance. Therefore the development of predictive tools such as the Mountain Contaminant Potential model (Wania and Westgate, 2008), may help to prevent pollution in mountain areas.

Suitable matrices for the detection of environmental contaminants may include mountain soils, lake sediments, snow, which may act as potential mountain ''reservoirs'' useful for monitoring purposes.. No generic media could be recommended as a monitoring tool, for any given location a careful revision of pros and cons must be conducted. In addition, many modern techniques are applied to observe POPs levels in different environmental matrices such air and sediments, which may be applied to monitor spatial and temporal trends of POPs in mountains areas around the world.

Given the concerns raised by scientists regarding the elevated levels of pollution observed in mountain areas, regulatory authorities and the general public must be called to attention of this problem, in order to establish proper protection measures of these important ecosystems throughout the world. We need to also consider that millions of people living in mountain areas may be subjected to increased pollution levels if a proper management is not conducted.

Acknowledgments

The authors thanks the support of FONDECYT n°1080294.

References

Barra, R., Popp, P., Quiroz, R., Bauer, C., Cid, H., Tumpling, W.v., 2005. Persistent toxic substances in soils and waters along an altitudinal gradient in the Laja River Basin, Central Southern Chile. *Chemosphere* 58, 905-915.

Barra, R., Popp, P., Quiroz, R., Treutler, H.-C., Araneda, A., Bauer, C., Urrutia, R., 2006. Polycyclic aromatic hydrocarbons fluxes during the past 50 years observed in dated sediment cores from Andean mountain lakes in central south Chile. *Ecotoxicology and Environmental Safety* 63, 52-60.

Blais, J.M., Schindler, D.W., Muir, D.C.G., Kimpe, L.E., Donald, D.B., Rosenberg, B., 1998. Accumulation of Persistent Organochlorine Compounds in Mountains of Western Canada. *Nature* 395, 585-588.

Borghini, F., Grimalt, J.O., Sanchez-Hernandez, J.C., Barra, R., García, C.J.T., Focardi, S., 2005. Organochlorine compounds in soils and sediments of the mountain Andean Lakes. *Environmental Pollution* 136, 253-266.

Calamari, D., Bacci, E., Focardi, S., Gaggi, C., Morosini, M., Vighi, M., 1991. Role of plant biomass in the global environmental partitioning of chlorinated hydrocarbons. *Environmental Science and Technology* 25, 1489-1495.

Camarero, L., Catalan, J., Pla, S., Rieradevall, M., Jiménez, M., Prat, N., Rodríguez, A., Encina, L., Cruz-Pizarro, L., Castillo, P.S., Carrillo, P., Toro, M., Grimalt, J., Berdie, L., Fernández, P., Vilanova, R., 1995. Remote mountain lakes as indicators of diffuse acidic and organic pollution in the Iberian peninsula (AL:PE 2 studies). Water, Air, andamp; *Soil Pollution* 85, 487-492.

Carrera, G., Fernandez, P., Vilanova, R.M., Grimalt, J.O., 2001. Persistent organic pollutants in snow from European high mountain areas. *Atmospheric Environment* 35, 245-254.

Daly, G.L., Wania, F., 2004. Simulating the Influence of Snow on the Fate of Organic Compounds. *Environmental science and technology* 38, 4176-4186.

Daly, G.L., Wania, F., 2005. Organic contaminants in mountains. *Environmental Science and Technology* 39, 385-398.

Datta, S., McConnell, L.L., Baker, J.E., Lenoir, J., Seiber, J.N., 1998. Evidence for Atmospheric Transport and Deposition of Polychlorinated Biphenyls to the Lake Tahoe Basin, California-Nevada. *Environmental Science and Technology* 32, 1378-1385.

Grimalt, J.O., Fernández, P., Berdié, L., Vilanova, R.M., Catalan, J., Psenner, R., Hofer, R., Appleby, P.G., Lien, L., Rosseland, B.O., Massabuau, J.-C., Battarbee, R.W., 2001. Selective Trapping of Organochlorine Compounds in Mountain Lakes of Temperate Areas. *Environmental Science and Technology* 35, 2690-2697.

Herbert, B.M.J., Halsall, C.J., Villa, S., Jones, K.C., Kallenborn, R., 2005. Rapid changes in PCB and OC pesticide concentrations in arctic snow. *Environmental Science and Technology* 39, 2998-3005.

Kang, J.-H., Choi, S.-D., Park, H., Baek, S.-Y., Hong, S., Chang, Y.-S., 2009. Atmospheric deposition of persistent organic pollutants to the East Rongbuk Glacier in the Himalayas. *Science of The Total Environment* 408, 57-63.

Muir, D.C.G., Grift, N.P., Lockhart, W.L., Wilkinson, P., Billeck, B.N., Brunskill, G.J., 1995. Spatial trends and historical profiles of organochlorine pesticides in Arctic lake sediments. *Science of The Total Environment* 160-161, 447-457.

Muir, D.C.G., Omelchenko, A., Grift, N.P., Savoie, D.A., Lockhart, W.L., Wilkinson, P., Brunskill, G.J., 1996. Spatial Trends and Historical Deposition of Polychlorinated Biphenyls in Canadian Midlatitude and Arctic Lake Sediments. *Environmental Science and Technology* 30, 3609-3617.

Pozo, K., Harner, T., Lee, S.C., Wania, F., Muir, D.C.G., Jones, K.C., 2008. Seasonally Resolved Concentrations of Persistent Organic Pollutants in the Global Atmosphere from the First Year of the GAPS Study. *Environmental science and technology* 43, 796-803.

Pozo, K., Harner, T., Wania, F., Muir, D.C.G., Jones, K.C., Barrie, L.A., 2006. Toward a Global Network for Persistent Organic Pollutants in Air:â€‰ Results from the GAPS Study. *Environmental science and technology* 40, 4867-4873.

Pozo, K., Urrutia, R., Barra, R., Mariottini, M., Treutler, H.-C., Araneda, A., Focardi, S., 2007. Records of polychlorinated biphenyls (PCBs) in sediments of four remote Chilean Andean Lakes. *Chemosphere* 66, 1911-1921.

Prévôt, A.S.H., Dommen, J., Bäumle, M., Furger, M., 2000. Diurnal variations of volatile organic compounds and local circulation systems in an Alpine valley. *Atmospheric Environment* 34, 1413-1423.

Quiroz, R., 2009. Compuestos Organicos Persistentes en Zonas de Alta Montaña. Departamento de Quimica Analitica. Universidad de Barcelona, Barcelona.

Quiroz, R., Arellano, L., Grimalt, J.O., Fernández, P., 2008. Analysis of polybrominated diphenyl ethers in atmospheric deposition and snow samples by solid-phase disk extraction. *Journal of Chromatography* A 1192, 147-151.

Quiroz, R., Popp, P., Barra, R., 2009. Analysis of PCB levels in snow from the Aconcagua Mountain (Southern Andes) using the stir bar sorptive extraction. *Environmental Chemistry Letters* 7, 283-288.

Sarkar, U.K., Basheer, V.S., Singh, A.K., Srivastava, S.M., 2003. Organochlorine Pesticide Residues in Water and Fish Samples: First Report from Rivers and Streams of Kumaon Himalayan Region, India. *Bulletin of Environmental Contamination and Toxicology* 70, 0485-0493.

Shoeib, M., Harner, T., 2002. Characterization and Comparison of Three Passive Air Samplers for Persistent Organic Pollutants. *Environmental science and technology* 36, 4142-4151.

UNEP, 2002. Mountain Watch Report UNEP World Conservation Monitoring Centre.

Vilanova, R.M., Fernández, P., Grimalt, J.O., 2001. Polychlorinated biphenyl partitioning in waters of a remote mountain lake. *The Science of the Total Environment* 279, 51-62.

Wang, X.-p., Gong, P., Yao, T.-d., Jones, K.C., 2010. Passive Air Sampling of Organochlorine Pesticides, Polychlorinated Biphenyls, and Polybrominated Diphenyl Ethers Across the Tibetan Plateau. *Environmental science and technology* 44, 2988-2993.

Wang, X.-p., Yao, T.-d., Cong, Z.-y., Yan, X.-l., Kang, S.-c., Zhang, Y., 2006. Gradient distribution of persistent organic contaminants along northern slope of central-Himalayas, China. *Science of The Total Environment* 372, 193-202.

Wania, F., 1997. Modelling the fate of non-polar organic chemicals in an ageing snow pack. *Chemosphere* 35, 2345-2363.

Wania, F., Hoff, J.T., Jia, C.Q., Mackay, D., 1998. The effects of snow and ice on the environmental behaviour of hydrophobic organic chemicals. *Environmental Pollution* 102, 25-41.

Wania, F.F., Mackay, D., Hoff, J.T., 1999. The importance of snow scavenging of polychlorinated biphenyl and polycyclic aromatic hydrocarbon vapors. *Environmental science and technology* 33, 195-197.

Wania F. , Westgate,J.F. 2008.On the mechanisms of the mountain cold trapping of organic chemicals. *Environmental Science and Technology* 42, 9092-9098

In: Mountain Ecosystems
Editor: Kevin E. Richards, pp. 93-100
ISBN 978-1-61209-306-2

Chapter 4

IMPACT OF LAND-USE CHANGE ON SEASONAL DYNAMICS OF TOTAL PROTEIN FLOW FROM ROOTS OF MOUNTAIN MEADOW PLANT COMMUNITIES

Valerie Vranova [1], Marian Pavelka [2], Klement Rejsek [1] and Pavel Formanek [1]

[1]Department of Geology and Soil Science,
Faculty of Forestry and Wood Technology,
Mendel University of Brno, Brno, Czech Republic
[2]Institute of Systems Biology and Ecology,
Laboratory of Plants Ecological Physiology, Brno,
Czech Republic

ABSTRACT

Socioeconomic changes in Central Europe have been accompanied by abandonment of previously managed meadows in mountain regions. Among others influences, alteration of N-cycling through such a land-use change may be reflected in N rhizodeposition including proteinaceous compounds. In this work we have determined no significant ($P>0.05$) effect of 15 year abandonment of previously long-term mown mountain meadows on total protein flow in the form of water-soluble root exudates in mineral soil of 3–8.5 cm depth. Throughout the vegetation season, 6–21 mg N per square meter was deposited every sampling occasion in exuded total proteins on both types of meadows.

Keywords: root exudates, proteins, meadows, management, mowing

INTRODUCTION

Recent socioeconomic changes have strongly influenced land use in mountain areas of Europe. This has recently led to setting up the EU project Carbomont, with the objective to examine the "effect of land use changes on sources, sinks and fluxes of carbon in European mountain areas" (EVK2-CT2001- 00125, EU, Framework 5). In the mountain regions of southern Europe, pastures that have degraded in the past due to combination of overgrazing and summer water shortage are currently recovering following a decrease in grazing pressure. On the other hand, a recent increase in reindeer stocks has increased grazing pressure in the subarctic tundra. In the Alps, a rapidly increasing proportion of formerly managed meadows and pastures has been abandoned over the recent decades and is now being recolonised by shrub and tree species. In the mountainous areas of the Central Europe, large areas have been abandoned following re-privatization of land.

Grassland ecosystems accumulate most of the primary production (60–90 %) underground (Stanton, 1988). Plant communities of these ecosystems are sensitive to different treatments including mowing or grazing. This sensitivity may explain the changes in rate of element cycling due to different reasons including alteration of quantity and quality of below-ground and above-ground plant biomass. Grassland use may also be reflected in C-and N-flow from roots into different soil horizons (Murray et al., 2004; Pinton et al., 2007). Water-soluble root exudates include compounds such as sugars, amino acids, organic acids, proteins, hormones, vitamins etc. (Pinton et al., 2007) where protein-N may represent a significant part of the total N of root exudates of different plant species representing a source for N-mineralization (Vancura and Kunc, 1988; Rejsek et al., 2008).

According to our best knowledge, there is little information on the effect of grassland-use change on N-cycling through flow of proteins in the form of water-soluble root exudates. Due to this we have attempted to determine experimentally the effect of abandonment of traditionally mown mountain meadows on protein exudation from plant roots in the course of one vegetation season.

MATERIALS AND METHODS

The experiments were performed near to the site "Bílý Kříž" in the Moravian-Silesian Beskids Mountains in the northeast part of the Czech Republic (N 49°30'17", E 18°32'28"), on a slope with an elevation of 825 – 860 m a. s. l. and a southeast orientation. The local subcontinental climate in this region is characterised by a mean annual air temperature of 4.9°C, a mean relative air humidity of 80 % and a mean annual precipitation of 1100 mm. There are 160 days with snow cover per year. The experimental meadow (1 ha) was originally mowed regularly, the hay removed and stored as feed for livestock. This traditional management practice ceased fifteen years ago on one half of the meadow (abandoned meadow), while the remaining half has been moderately mown once a season in July or August. Before abandonment, the original area of two study plots was uniform as regards the plant populations and soil properties (Formanek et al., 2008a). Nowadays, the moderately mown meadow plant community belongs to the *Nardo-Callunetea* class while the abandoned meadow is covered with the *Molinio-Arrhenatheretea* class plant community (Zelena in Formanek et al., 2008a). The soil of both meadows is classified as Gleyic Luvisol (ISSS-ISRIC-FAO 1998) (see Table 1).

Root samples were collected from the abandoned and the moderately mown mountain meadows in approx. 30-day intervals throughout the vegetation season 2008. A metal cylinder (diameter 10cm) was inserted to a depth of 8.5 cm at five randomly selected points within each of the two treatment areas. After transport, the samples were stored overnight at 6 °C. Aboveground biomass with 3 cm thick organic layer was cut-off and roots were separated from thicker mineral soil. Soil was carefully washed off and roots were rinsed several times with tap water and, consequently, with demineralized water. A disadvantage of this procedure is damage to the roots due to sampling and cutting. Root exudates were collected in 500 ml of 0.01 M $CaSO_4$ for 2 hours at room temperature. Afterwards, the collection medium was filtered through paper and then through 0.45 μm membrane filter. The total proteins were assessed according to Bradford (1976) using bovine serum albumin (BSA) as the standard.

Table 1. Properties of soil layer where exudation of total proteins from plant roots on moderately mown and abandoned mountain meadows was determined (Mean ± SD; n = 3-8). The different letters mark significant (P<0.05) differences between the meadows

Soil properties	Moderately mown	Abandoned meadow
Clay (%)	18.60 ± 3.4^a	19.93 ± 2.3^a
Silt (%)	26.0 ± 3.7^a	27.1 ± 7.0^a
Sand (%)	55.40 ± 6.5^a	52.93 ± 9.1^a
pH/H_2O	4.30 ± 0.45^a	4.27 ± 0.42^a
pH/0.01 M $CaCl_2$	3.80 ± 0.19^a	3.78 ± 0.18^a
CEC ($mmol_c\,kg^{-1}$)	159.34 ± 2.28^a	184.85 ± 10.01^b
C_t (%)	4.76 ± 0.76^a	4.77 ± 0.71^a
N_t (%)	0.29 ± 0.06^a	0.33 ± 0.09^a
C/N	16.91 ± 3.58^a	15.24 ± 3.14^a
C-HA (%)	0.50 ± 0.05^a	0.57 ± 0.11^a
C-FA (%)	0.72 ± 0.05^a	0.77 ± 0.05^a

Formanek et al., 2008b.

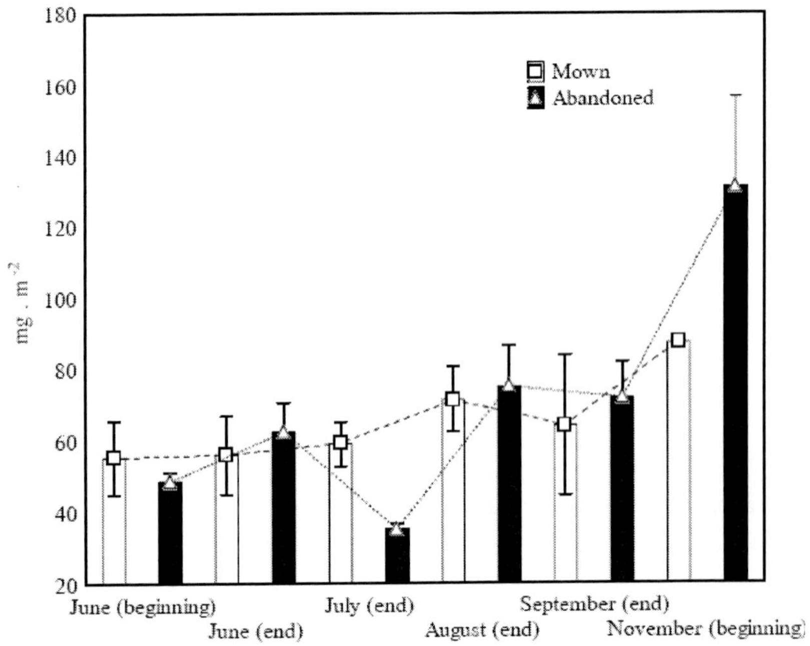

Figure 1. Amount of proteins in root exudates of plant community of differently managed mountain meadows (Mean ± SE).

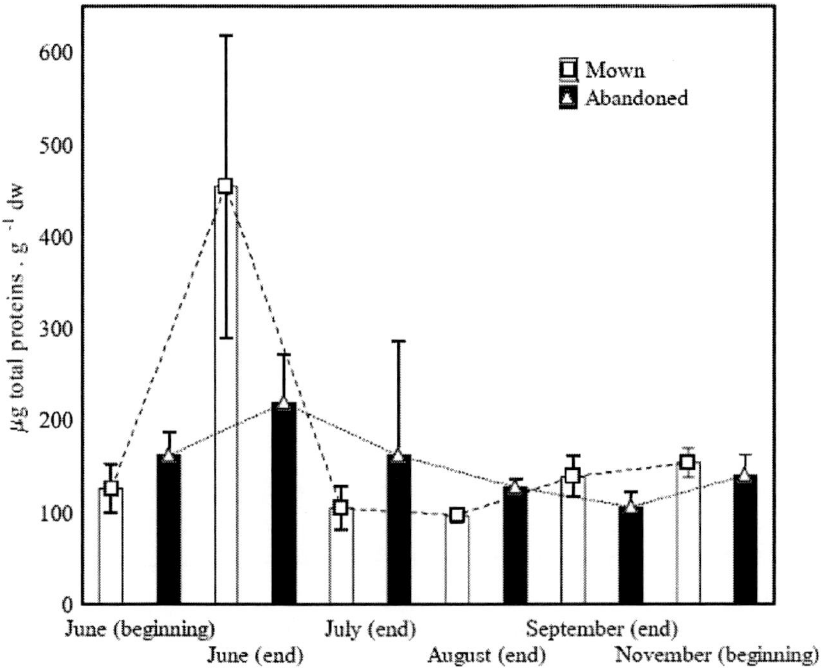

Figure 2. Amount of proteins in plant root exudates released per g dry root biomass (Mean ± SE).

Table 2. Relative heterogeneity (CV, %) of total proteins exudation per square meter on abandoned and moderately mown meadows

	June (beg)	June (end)	July (end)	August (end)	September (end)	November (beg)
Mown	37.8	35.0	20.5	28.7	36.2	3.4
Abandoned	11.4	29.8	6.1	34.6	31.6	38.5

As the preconditions for parametric statistical testing were not fulfilled even after log-transformation of data, nonparametric Kruskal-Wallis ANOVAs *plus* Dannet's *t*-tests were used (Statistica 8.0).

Results and Discussion

Exudation of total proteins from roots of differently managed meadow plant communities did not differ significantly (P>0.05) and the amount of total proteins released per square meter ranged from 35 to 131 mg representing approx. 6–21 mg of N. Protein exudation is known to differ according to plant species and plant age. For example, *Medicago sativa* was found to secrete 7.7 times more proteins than *Arabidopsis thaliana* (De-la-Peňa et al., 2008). Our results indicate that dynamics of different plant species occurrence and growth on the studied meadows probably masked the differences between the differently managed ecosystems. The amount of total proteins released per surface area increased throughout the vegetation season with only one marked decline at the end of July on the abandoned meadow (Figure 1). Contrary to these results, the total proteins in root exudates per g of dry root biomass were the highest in the first part of the vegetation season (Figure 2). Root exudation related to weight of root biomass is known to decrease with plant age probably due to decreasing metabolic activity of superficial root cells (Hamlen et al., 1972; Pinton et al., 2007). Mowing performed just at the day of sampling (end of July) had no effect on protein exudation related to unit of dry root biomass, which is in agreement with the experiments of e.g. Kuzyakov et al., (2002) on soil casein-protease activity. Relative stand heterogeneity of total protein exudation per surface unit ranged from 3.4 to 38.5 % throughout the season (Table 2) and was considerably lower compared to concentrations of bioavailable N-forms (Formanek et al., 2008b).

Conclusion

Although N rhizodeposition has been thought to have a significant role in N cycling and rhizosphere N dynamics, modification of these processes due to grassland-use change was not proved in this study when the release of total proteins in water-soluble root exudates was considered. Further research taking into account other N-containing compounds, especially amino acids, separating their efflux and re-uptake by plant roots is recomended to better evaluate the effect of grassland management on N-input in the form of water-soluble root exudates.

Acknowledgments

The study was supported by the Project No. VZ MSM 6215648902, part 4/2/2, and the Project "An assessment of the effect of management of mountain meadows on rate of release of total water-soluble carbon, carbohydrates and proteins from plant roots (DC)".

Peer-Reviewer

Dr. Keith R. Skene, Lecturer in Plant Ecology from the University of Dundee, Department of Biological Sciences, Dundee DD1 4HN, the United Kingdom, telephone: (0)1382 344257, fax: (0)1382 344275, E-mail: k.r.skene@dundee.ac.uk.

References

Bradford, M. M., (1976). A rapid and sensitive method for the quantitation of micro-gram quantities of protein utilising the principle of protein-dye binding. *Analytical Biochemistry, 72*, 248–254.

De-la-Peňa, C., Lei, Z., Watson, B. S., Summer, L. W. and Vivanco, J. M. (2008). Root-microbe communication through protein secretion. *Journal of Biological Chemistry, 283*, 25247–25255.

Formanek, P., Rejsek, K., Vranova, V., and Marek, M. V. (2008a). Bioavailable amino acids and mineral nitrogen forms in soil of moderately mown and abandoned mountain meadows. *Amino Acids, 34*, 301–306.

Formanek, P., Rejsek, K., Vranova, V., and Klejdus, B. (2008b). Selected diamino acids in soils of differently managed mountain meadow and forest ecosystems: assessment of their role in ecosystem nutrition. In: Schäfer, H. A. and Wohlbier, L. M. (Eds.) *Diamino amino acids*, pp. 183–217. New York, US: Nova Science Publishers.

Hamlen, R. A., Lukezic, F. L. and Bloom, J. R. (1972). Influence of age and stage of development on the neutral carbohydrate components in root exudates from alfalfa plants grown in a gnotobiotic environment. *Canadian Journal of Plant Science,52,* 633–642.

ISSS-ISRIC-FAO (1998). World reference basis for soil resources. FAO, Rome.

Kuzyakov, Y., Biryukova, O. V., Kaznetzova, T. V., Mölter, K., Kandeler, E. and Stahr, K. (2002). Carbon partitioning in plant and soil, carbon dioxide fluxes and enzyme activities as affected by cutting ryegrass. *Biology and Fertility of Soils, 35*, 348–358.

Murray, P., Ostle, N., Kenny, C. and Grant, H. (2004). Effect of defoliation on patterns of carbon exudation from *Agrostis capillaris*. *Plant Nutrition and Soil Science 167*, 487–493.

Pinton, R., Varanini, Z. and Nannipieri, P. (2007). *The rhizosphere: biochemistry and organic substances at the soil-plant interface.* Boca Raton, US: Taylor and Francis Group.

Rejsek, K., Formanek, P., and Pavelka, M. (2008). Estimation of protease activity in soils at low temperatures by casein amendment and with substitution of buffer by demineralized water. *Amino Acids, 35*, 411–417.

Santon, N. L. (1988). The underground in grasslands. *Annual Review of Ecology and Systematics, 19*, 573–589.

Vancura, V., Kunc, F. (1988). *Soil microbial associations: control of structures and functions*, Praha, Czech Republic: Academia.

In: Mountain Ecosystems ISBN 978-1-61209-306-2
Editor: Kevin E. Richards, pp. 101-121

Chapter 5

ATMOSPHERIC CARBON DIOXIDE TRANSPORT OVER MOUNTAINOUS TERRAIN

Jielun Sun*
National Center for Atmospheric Research, Boulder, Colorado, USA
Stephan F.J. De Wekker
University of Virginia, Charlottesville, Virginia, USA

Abstract

We investigated trace gas transport by daytime mountain-induced circulations based on aircraft data collected during the Airborne Carbon in the Mountains Experiment (ACME04) conducted in 2004. Using measurements of traditional meteorology and trace gases associated with ecosystem processes, we found evidence of thermally-induced vertical mountain circulations consisting of an upslope flow, a horizontal return flow, and a descending flow along the Front Range of Colorado. Turbulent air was generated by convective buoyancy along the mountain slopes and ridges and by strong shear at the mountain ridge tops. The deep turbulent mixed layer was subsequently advected by the ambient large-scale and return flows across the plains east of the Front Range. The descending flow over the plains increased the atmospheric stability, which reduced the vertical mixing above the convective boundary layer over the plains. The mountain circulation and its interaction with the ambient flow and convective turbulent mixing effectively transported the air with its unique

*E-mail address: jsun@ucar.edu (Corresponding author)

characteristics of the CO_2, water vapor, and CO concentrations at the bottom of the atmosphere over the plains to above the convective boundary layer over the plains. This study suggest that mountain-induced circulations can have significant impacts on estimates of the regional ecosystem-atmosphere carbon exchange.

Keywords: atmospheric boundary layer, carbon dioxide, complex terrain, mountain circulation

1. Introduction

Significant progress has been made in investigating mesoscale circulations in complex terrain and their importance in weather forecasting, cloud formation, aviation, and air pollution transport (e.g., Geiger et al., 1995; Whiteman, 2000; Rotach and Zardi, 2007). However, turbulence characteristics of mountain-induced circulations and the role of the mountain-induced circulation in transporting biogenic and anthropogenic trace gases have not been fully investigated. The effect of mountains on the atmosphere is especially important around the Colorado Rocky Mountains, which is evident from the frequent occurrence of turbulent events encountered by commercial aircraft (Wolff and Sherman, 2008).

Air flows can be influenced by two types of mountain forcing. The mechanical forcing of the air flow over and around mountains has been studied in field experiments, in wind tunnels, and with numerical models (e.g., Blumen, 1990). The air flow over mountain tops can be compressed and the air speed enhanced, which leads to turbulence generation. Air flows over mountainous terrain can also be influenced by daytime solar heating and nighttime longwave radiative cooling of mountains (Whiteman, 2000). Because of the solar heating of mountains during the day, the air temperature over the mountain is warmer than the air at the same altitude above the plains that surround the mountain. The difference between the vertically-integrated temperature over the mountain and the plains leads to a horizontal gradient of the atmospheric pressure, which drives the air towards the mountain and leads to an upslope flow. The forcing of the air flow is related to the thermal difference of the mountain and the air at the same altitude, and is referred to as thermal forcing, in contrast to the mechanical forcing mentioned earlier. At night, the radiative cooling of the mountain reverses the thermal forcing and leads to a downslope flow. Both mechanical and thermal forcing can influence the air flow over the mountain, and the latter has a distinctive diurnal cycle.

The nighttime downslope flow has been studied in great detail. For example, the influence of the ambient large-scale flow on the nocturnal downslope flow was studied during the Atmospheric Studies in Complex Terrain (ASCOT) in the early 1980s (Barr and Orgill, 1989). The nighttime downslope flow occurs not only over steep valleys (Orgill et al., 1992) but also over seemingly flat terrain (Mahrt and Larsen, 1990). The downslope flow transports cold air downward, leading to stably stratified cold air pools and suppressed turbulent mixing over the plains. Recently, the important role of the nighttime downslope flow in transporting ecosystem respired CO_2 and ground-evaporated moisture has been investigated (e.g., Sun et al., 2007; Kutsch et al., 2008; Aubinet, 2008; Sun et al., 2010). As a result of the nighttime downslope flow and cold-air pooling, the air over the plains is characterized by cold, moist, and CO_2-rich air as well as by pollutants, such as CO and aerosols, if the plains are urbanized.

As the thermal forcing of complex terrain reverses after sunrise, the upslope flow induces a vertical mountain circulation. For a two-dimensional mountain with a long mountain ridge and a weak ambient flow perpendicular to the mountain ridge, the thermally-forced mountain circulation (hereafter, mountain circulation) is mainly in the vertical plane. Bounded by mass conservation, the daytime mountain circulation theoretically consists of an upslope flow, a return flow, which is the upslope flow turning away from the mountain, and a descending flow over the plains (e.g., Geiger et al., 1995). If the ambient wind is strong, it can flow over the mountain ridge to the lee side of the mountain. In this situation, the mountain circulation can be confined below the mountain ridge top or completely embedded in the dominant ambient wind. In reality, interactions among the large-scale ambient flow, the upslope flow, and the convective boundary layer over the plains make the observation of the return flow difficult (e.g., McGowan, 2004; Reuten et al., 2005).

The development of the upslope flow depends on several factors, such as the orientation of the slope, the incident solar radiation, and the strength of the ambient wind. The upslope flow has been found to begin one to two hours after sunrise (Orgill, 1989; Whiteman, 1989). Because of the stronger stratification in the cold air pool over the plains compared to that over the slope, the morning development of the convective boundary layer (CBL) over the slope and the mountain ridge is faster than the development over the plains (e.g., Banta, 1984; 1986; Rotach et al., 2004). Therefore, cold air, trace gases, and pollutants at the bottom of the atmosphere over the plains can be transported upward by both the upslope flow and the vertical convective turbulent transport (e.g., Gudiksen and Shearer, 1989; Kalthoff et al., 2000). The upslope flow leads to an efficient upward transport of moisture, pollutants and aerosols (e.g., Banta, 1984; McKendry et al., 1998; De Wekker et al., 2004). Interactions among the mountain circulation, the ambient flow, and the turbulent mixing can also lead to a deep turbulent mixed layer above the mountain ridge. When advected downwind from the mountain by the ambient and return flows, this layer frequently transitions to an elevated mixed layer over the plains (e.g., Arritt et al., 1992). The vertical coupling between the elevated turbulent layer and the CBL over the plains can lead to a rapid breakup of the cold air pool (e.g., Banta, 1984; 1985; 1986; Whiteman, 1982a & b). On the other hand, the descending flow of the mountain circulation can lead to enhanced stratification above the plains (e.g., De Wekker, 2008). In contrast to the nighttime downslope flow in transporting the ecosystem-respired CO_2, the daytime mountain circulation can transport the nighttime-accumulated CO_2-rich air upslope (Sun et al., 2010).

Direct observations of interactions between mountain circulations and convective turbulence over sunny slopes as well as the vertical transport of CO_2 by mountain circulations, especially airborne observations, have been sparse. Recent examples include two CARBOEUROPE airborne field campaigns, one in Southwestern France (Dolman et al., 2006; Ahmadov et al., 2007) and one in Spain (Pérez-Landa et al., 2007a & b), which were conducted to estimate the regional carbon balance. In this study, we focus on aircraft observations over the Front Range of the Colorado Rocky mountains to investigate the role of the daytime mountain circulation in transporting trace gases, such as CO_2, CO, and aerosols. Our unique aircraft observations include a large number of soundings, turbulence measurements, and a suite of trace gas measurements. The trace gas measurements allow a systematic investigation of 1) dynamic aspects of mountain circulations by using trace gasses as tracers, and 2) trace gas transport by mountain circulations and turbulent mixing under a variety of ambient wind and stability conditions. We summarize the dataset used in the study in section 2. In section 3, we focus on characteristics of the daytime mountain circulation and use trace gases, such as CO_2, to identify the mountain circulation. In addition, we investigate the spatial and diurnal variation of the CO_2 concentration as a result of the mountain circulation. A summary and the implication of the observed transport of CO_2 by the mountain circulation on the regional CO_2 budget are given in section 4.

2. Observations

The data used in this study were collected during the Airborne Carbon in the Mountains Experiment (ACME04) in 2004 (Sun et al., 2010). A total of 16 flights were conducted with the C-130 aircraft from the National Center for Atmospheric Research (NCAR). In general, four flight patterns were flown during ACME04: 1) the racetrack, 2) the roller coaster, 3) the slant and spiral sounding, and 4) the Hayman fire. In this study, we concentrate on the flight patterns of the roller coaster and the spiral runs over the Front Range of the Colorado Rocky Mountains (Fig. 1). The roller coaster run consisted of a terrain following flight across the eastern slope of the Front Range (blue track in Fig. 1), which was commonly combined with the spiral sounding run (orange track in Fig. 1) located $\sim$ 35 km east of the Front Range mountain ridges. Sometimes, a level run at $\sim$ 5000 m altitude was added to connect the roller coaster and the spiral runs.

Wind, temperature, and water vapor were measured by fast response sensors on the aircraft and sampled at 25 samples per second, the CO_2 and CO concentrations at five samples per second, and O_3 and aerosol concentrations at one sample per second. The variability of the data from the fast-response sensors can be used to identify turbulent air. The zero value of the vertical velocity is within the calibration uncertainty; therefore, its fluctuation, not its mean value, was used in the study. Potential temperature, which is conserved as an air parcel moves vertically without any energy exchange with its environment, was calculated from the observed air temperature (e.g., see Wallace and Hobbs, 2006). Water vapor specific humidity, which is the mass of the water vapor for a given mass of air, was used to describe the moisture content of the atmosphere. To ensure the accuracy of the CO_2 concentration measurements, an on-board calibration procedure was carried out frequently during each flight, resulting in data gaps. During ACME04, a ground-based field campaign named the Carbon in the Mountains Experiment (CME04) was conducted at the Niwot Ridge Long-Term Ecological Reserve (LTER) site, which is also an AmeriFlux site in Colorado. The site is located directly under the roller coaster track (marked as the triangle in Fig. 1). All the aircraft and ground measurements and their accuracies were described by Sun et al. (2010).

We analyzed two flight days in this study: the morning research flight 11 (RF11) on 22 July, and the morning and afternoon research flights, RF12, and RF13, on 26 July. The weather conditions during those flights were mostly clear to partly cloudy with relatively weak ambient westerlies (Fig. 2). A significant

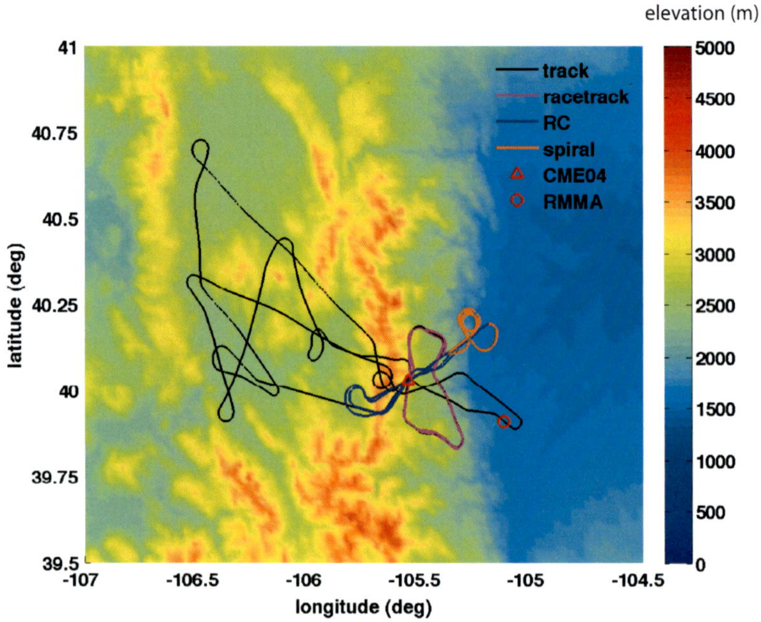

Figure 1. Flight track of RF12, which shows the racetrack, the roller coaster and the spiral tracks, and the location of CME04. The NCAR C-130 aircraft operated from the Rocky Mountain Metropolitan Airport (RMMA) during ACME04.

two-day rain event occurred after RF11. The sunny weather, which occurred the day before RF12, led to large downward CO_2 fluxes. We use local standard time (LST), which is seven hours behind Coordinated Universal Time (UTC). The vertical coordinate used in this study is the pressure altitude above sea level (ASL).

3. Characteristics of Daytime Mountain Circulations and Their Transport of CO_2

The flights over the Front Range allow an investigation of the dynamics of the daytime mountain circulation and trace gas transport by the mountain circulation. We first focus on RF11 to investigate the morning mountain circulation. Then we examine the temporal variation of the mountain circulation by ana-

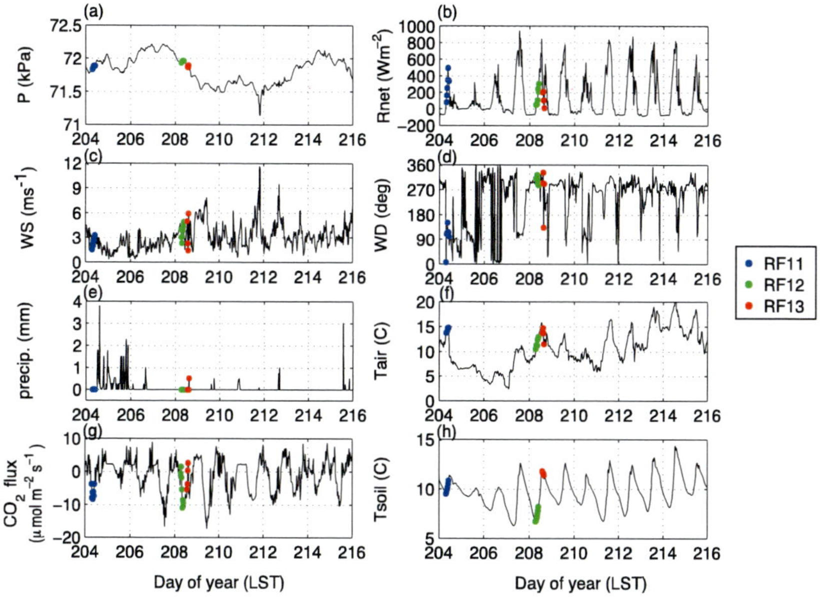

Figure 2. The time series of (a) air pressure (P) at 12 m above the ground, (b) net radiation (Rnet) at 25 m, (c) wind speed (WS), and (d) direction (WD) at 21 m, (e) precipitation at 10.5 m, (f) air temperature (Tair), (g) CO_2 flux at 21 m, and (h) soil temperature (Tsoil) at 5 cm depth at one of the CME04 towers operated by the University of Colorado. The three flights that were analyzed in this study are marked with three different colors on each time series.

lyzing the observations from RF12 and RF13 that were conducted on the same day.

3.1. 22 July

In the early morning of RF11, the aircraft descended along the sunny slope following the roller coaster track, spiraled up to about 5000 m, and flew southwestward along the roller coaster track at 5000 m (Figs. 3a and b). The air over the plains was relatively stable, which was evident in the increasing potential temperature with height (Fig. 3f), and the decreasing water vapor and CO_2 concentrations with height (Figs. 3g and h). The mountain ridge height along the Front Range is around 3600 m. It peaks at $\sim$ 3450 m along the roller coaster track (marked z_{top} in Fig.3), and $\sim$ 3900 m west of the spiral run. The relatively high water vapor and CO_2 concentrations at the bottom of the atmosphere over the plains are due to the combined effect of the ground evaporation and ecosystem respiration during the previous night and the weak vertical turbulent transport as a result of the stably stratified nocturnal atmosphere boundary layer. Ground heating after sunrise at 0449 LST produced convective turbulent mixing along the sunny slope as evident from CME04 (Sun et al., 2010; Burns et al., 2011). The relatively low turbulence strength as demonstrated in weak fluctuations of the vertical velocity far below the mountain ridge (Fig.3e) indicates that the CBL along the sunny slope was below the aircraft flight level at the time. The upper air over the plains was less turbulent than the air over the sunny slope at the same altitude.

The daytime thermally-forced upslope flow, opposite from the ambient westerlies, was evident from the easterly wind below $\sim$ 2900 m along the roller coaster run (marked z^b_{EML} in Fig. 3) and below $\sim$ 3100 m from the spiral run (Fig. 3d). The difference in the depth of the easterly flow at the two locations probably results from the spatial variation of the combined large-scale pressure forcing and the thermal forcing associated with the variation of the Front Range mountain ridge heights. As the upslope flow encountered the ambient air flowing down the mountain ridge along the sunny slope at z^b_{EML} in Fig. 3, the convergence led to the eastward transport of the convective turbulence that had developed along the sunny slope. Because the eastward return flow was in the same direction as the ambient wind in this case, the return flow appeared as an enhancement of the wind speed just above z^b_{EML} (Fig. 3c). In addition, the enhancement of the water vapor and CO concentrations from the upslope transport of the plains air was also observed in Figs. 3g and i. Because of the existence of the urban area at the Front Range, the observed high CO concentration (Fig. 3i) was likely associated with anthropogenic pollution (Seinfeld and Pandis, 2006). The simultaneous increase of the water vapor, the CO concentration, and the wind speed above z^b_{EML} confirmed the presence of the return flow as part of the thermally-forced vertical circulation, which is schematically shown in Fig. 4. If the return flow carried part of the plains air, we would expect the CO_2 concentration to increase above z^b_{EML}, which was not clearly evident in Fig. 3h. The absence of the clear CO_2 enhancement is partly due to the missing CO_2 observation below z^b_{EML} because of the CO_2 sensor calibration. In addition, as the boundary layer air over the plains was transported upward by the upslope flow along the sunny slope, the high CO_2 concentration could be reduced by

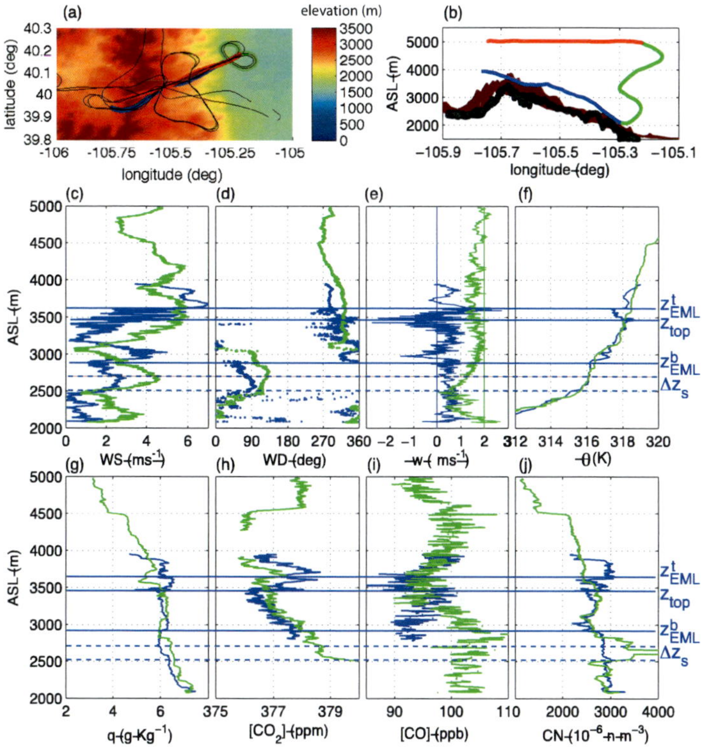

Figure 3. The aircraft observations along the roller coaster run (blue), and the spiral run (green) during RF11. The two runs were flown consecutively at 0622 LST and 0627 LST. Their flight tracks and the topography are in (a). Their altitudes are in (b), where the topography under the roller coaster run is in black, and the topography west of the spiral run is in brown. The observations include (c) wind speed, (d) wind direction, (e) vertical velocity, (f) potential temperature, (g) water vapor specific humidity, and the concentrations of (h) CO_2, (i) CO, and (j) aerosols as functions of altitude. For easy comparison, the vertical velocity for the spiral run is shifted by 2 ms^{-1}, and the zero vertical velocity for each run is marked in its track color in (e). The approximate bottom and top of the elevated mixed layer (EML) for the roller coaster run are marked with the horizontal lines and labeled z^b_{EML} and z^t_{EML}, respectively. The mountain top under the roller coaster run is marked z_{top}. The relatively stable layer, Δz_s, as a result of the descending flow of the mountain circulation, is marked with the dashed lines.

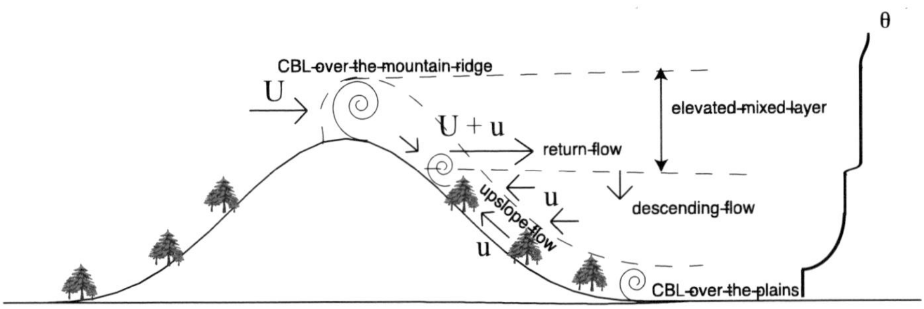

Figure 4. Schematic summary of the morning mountain circulation and the simplified potential temperature (θ) profile over the plains downstream from the mountain from RF11. Here, U and u represent the ambient westerly flow, and the thermally driven upslope flow. The spiral symbols represent turbulent air. The dashed line along the slope represents the top of the convective boundary layer (CBL). The approximately horizontal dashed lines represent the top and the bottom of the elevated mixed layer (EML).

both photosynthetic uptake of mountain vegetation and the upward convective turbulent mixing along the slope, and were observed on the slope surface at CME04 (Sun et al., 2010; Burns et al., 2011). By contrast, water vapor release as a by-product of photosynthesis as well as soil evaporation can enhance the air water vapor content. The upslope transport of the air along the eastern slope of the Front Range was also evident from the carbon isotope analysis by comparing the air samples collected by the aircraft and at the CME04 site (Sun et al., 2010). Because of the uncertainty of the zero velocity calibration, the descending portion of the mountain circulation cannot be directly measured by the aircraft. However, its effect in stabilizing the atmosphere was visible between ~ 2500 m and ~ 2750 m from the roller coaster run (marked by Δz_s between the dashed lines in Fig. 3) in 1) the reduced high-frequency oscillation of the vertical velocity, 2) the increased air temperature, 3) the reduced water vapor, and 4) the increased wind speed as a result of the reduced turbulent mixing.

Near the mountain ridge top (z_{top} in Fig. 3), the strong wind shear associated with the strong westerly ambient flow and buoyant turbulence led to the strongly turbulent CBL in the layer that was slightly below z_{top} and z^t_{EML}

in Fig. 3e. As a result of the turbulent mixing within the CBL, the potential temperature and the water vapor specific humidity were almost invariant with height in this layer (Figs. 3f and g). The higher concentrations of CO_2, water vapor, and aerosols, and the lower CO concentration within the CBL compared to those below the mountain ridge top (Figs. 3g, h, and j) could be natural and anthropogenic. Naturally, the air close to the mountain ridge top could be related to the soil respiration and evaporation above the tree line of $\sim$ 3300 m, and possible CO deposition to soil surfaces at night (e.g. Conrad and Seiler, 1985). Anthropogenically, the efficient power plant at Hayden, Colorado, upstream of the roller coaster run could also lead to the high CO_2, water vapor, and aerosol, and low CO concentrations. Because of the magnitude of the CO_2 and CO change with height, the observed high-CO_2 and low-CO air above z_{top} is more likely associated with the power plant upstream. The sharp reduction of the turbulence above $\sim$ 3600 m (marked z_{EML}^t in Fig. 3) could be due to lack of buoyancy either above the top of the CBL over the mountain ridge top or over the shady west side of the mountain as the aircraft flew across the sunny and shady sides of the mountains below the top of the CBL over the mountain ridge (see Fig. 4 for the schematic). If the latter was true, the depth of the turbulent layer over the sunny side of the mountains would be higher than z_{EML}^t.

Downstream of the Front Range mountains, turbulence was enhanced above the elevation of the convergence zone on the sunny slope, and was further enhanced close to the mountain ridge top. Advected by the ambient and the return flows, the deep turbulent layer was extended downstream from the mountains as an elevated mixed layer (EML) above the relatively stable upper-level air over the plains. The dynamical and ecological processes contributed to the variability in CO_2, CO, water vapor, and aerosol concentrations, and the formation of the sublayers downstream from the mountains. The observed EML from RF11 was turbulent but not well mixed in the vertical, which is similar to the results from the numerical simulations over the Front Range by Arritt et al. (1992).

Due to turbulence energy dissipation, the turbulence intensity decreased with distance from the Front Range mountains. However, the EML was still visible in the vertical velocity oscillation from the spiral run between $\sim z_{EML}^b$ and 4000 m (Fig. 3e). The higher top of the turbulent layer ($\sim$ 4000 m) compared to the top of the turbulent layer from the roller coaster run (z_{EML}^t in Fig. 3) is more likely due to the higher mountain ridge tops upwind of the spiral run compared to those along the roller coaster run (Fig. 3b). The relatively low CO_2 and high CO concentrations above 4000 m observed from the spiral run may represent the previous daytime air from a few hundred of kms upstream transported by the ambient westerly flow during the previous night. The low CO_2 concentration could result from the combined effect of ecosystem photosynthesis and convective mixing over a deep atmospheric boundary layer. The high CO concentration could be due to anthropogenic emissions, possible biogenic photoproduction (e.g., Tarr et al., 1995), and abiotic processes associated with chemical oxidation of soil organic matter (e.g., Conrad and Seiler, 1985). Figure 4 shows a schematic summary of the important processes during the morning mountain circulation that were derived from the RF11 analyses.

3.2. 26 July

The temporal variation of the mountain circulation from morning to early afternoon was investigated by analyzing two sets of the roller coaster and spiral runs from RF12 and one set from RF13. The three roller coaster runs were flown at $\sim$ 0733, 0839, and 1437 LST (Fig. 5), and the three spiral runs were flown at $\sim$ 0740, 0844, and 1442 LST (Fig. 6).

At the time of the first roller coaster and spiral runs (blue traces in Figs. 5 and 6), a stably-stratified air layer below the convergence zone (blue z^b_{EML} in Fig. 5) was observed in both the potential temperature (Figs. 5f and 6f) and the vertical velocity (Figs. 5e and 6e), which was similar to the air flow during RF11. Compared to RF11, the stronger ambient northwesterly flow resulted in the northwesterly flow along the roller coaster run down to $\sim$ 2450 m (marked with blue z^b_{EML} in Fig. 5), which was about 400-500 m lower than the one from RF11. The ambient flow was observed above $\sim$ 2500 m along the spiral run (marked blue z^b_{EML} in Fig. 6). The flow at the bottom of the atmosphere over the plains was influenced primarily by a high-pressure system southeast of the flight location as well as by thermal forcing of the mountains; therefore, it was southerly, instead of easterly, for the two morning runs. The differences in the CO_2, CO, and water vapor concentrations above and below the altitude where the ambient westerly flow and the southerly flow converged (blue z^b_{EML} in both Figs. 5 and 6) were due to the characteristics of the different air masses. Again, the high CO_2 and aerosol, and low CO air was observed close to the mountain ridge top (between blue z_m and z^t_{EML} in Fig. 5), which indicate the possible effect of the power plant upstream for the roller coaster run as explained in the previous subsection. The high CO_2 and relatively high CO and aerosol air above the blue z^t_{EML} may reflect the previous daytime CBL air upstream.

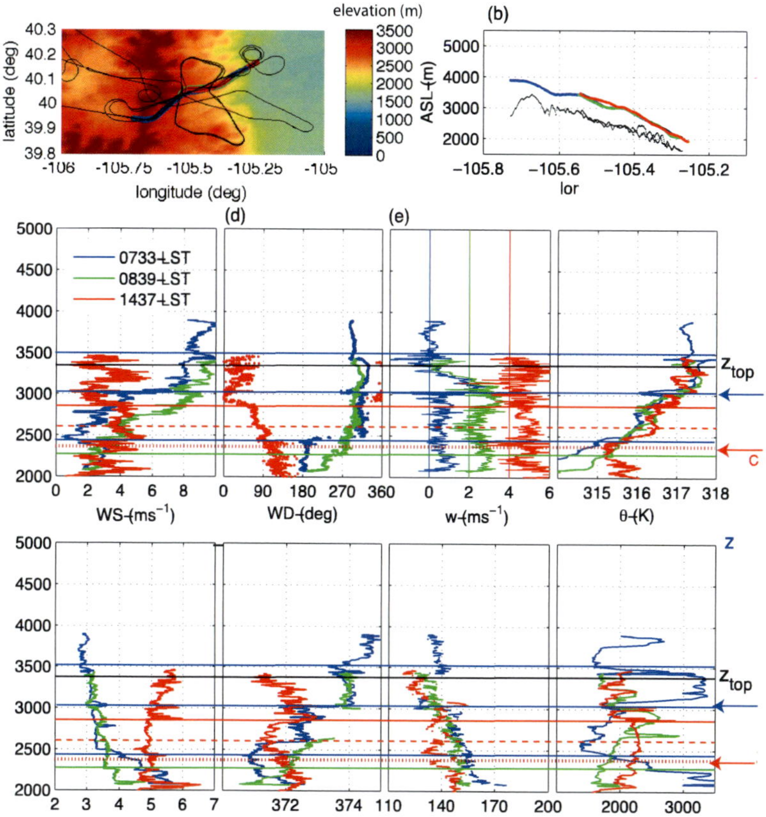

Figure 5. The aircraft observation of (c) wind speed, (d) wind direction, (e) vertical velocity, (f) potential temperature, (g) water vapor specific humidity, and the concentrations of (h) CO_2, (i) CO, and (j) aerosol as functions of altitude along the roller coaster run at 0733 LST (blue) and 0839 LST (green) from RF12, and at 1437 LST (red) from RF13. The flight tracks are marked in (a) and their altitudes are in (b), where the topography is colored in (a) and in black in (b). The top and bottom of the EML for each run are marked with z^t_{EML} and z^b_{EML} in the color of the track, respectively. Because the z^t_{EML} at 0839 LST and 1437 LST were above the flight track, no green and red z^t_{EML} are in this plot. The vertical layer between the red z^b_{EML} and the red dashed line marks the relatively stable layer associated with the descending flow for that run. The red dotted line marks the top of the CBL over the plains at the time of RF13. The black line labeled with z_{top} marks the top of the mountain ridge along the roller coaster track. The blue z_m marks the bottom of the ambient air flow that was not strongly influenced by the eastern slope.

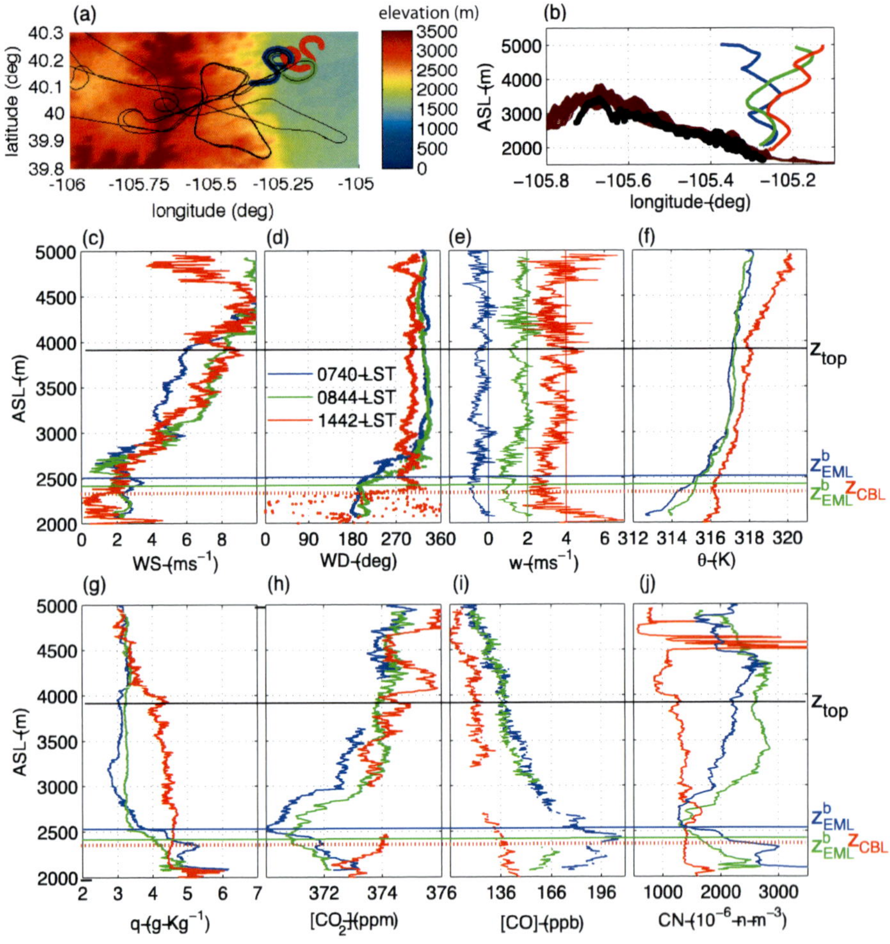

Figure 6. The same as Fig. 5 except for the three spiral runs east of the Front Range. The topography upwind of the spiral run and along the roller coaster run are marked in brown and black in (b), respectively. Here, z_{top} marks the approximate mountain ridge top upwind of the spiral run.

At 0839 LST (green traces in Figs. 5 and 6), the northwesterly ambient wind strengthened. The convergence between the ambient northwesterly and the southerly flows propagated down along the sunny slope (marked green z^b_{EML} in Figs. 5d and 6d). The downslope movement of the ambient northwesterly flow was also visible in 1) the enhanced turbulence, 2) the increased potential temperature, and 3) the increased CO_2 concentrations from the 0839 LST run between the blue and green z^b_{EML} along both roller-coaster and spiral runs (Figs. 5 and 6).

After refueling, the aircraft went back and flew another set of the roller coaster and spiral runs (red traces in Figs. 5 and 6). By then the air had become strongly turbulent up to $\sim$ 2400 m (marked red dotted line with z_{CBL} in both Figs. 5 and 6) over both the eastern slope and the plains. Along the roller coaster run, the evaporation, ecosystem photosynthesis, and the turbulent mixing led to warming, increased water vapor content, and decreased CO_2 concentration over the plains CBL. The flow from the ground up to $\sim$ 2800 m along the eastern slope (marked red z^b_{EML} in Fig. 5) changed from the morning southerly to the easterly upslope direction. In contrast, the wind direction over the plains was about the same as the ambient northwesterly wind all the way down to the top of the CBL (marked with red z_{CBL} in Fig. 6). The absence of the upslope flow at the location of the spiral run indicates that the influence of the thermal forcing of the Front Range on the air flow was limited to the area close to the Front Range. The reduced thermal forcing could be associated with the cloud development as shown in the reduced net radiation and reduced CO_2 fluxes at CME04 during the time of the afternoon flight (red marks in Figs. 2b and g). As a result of the reduced CO_2 uptake by vegetation and the reduced vertical turbulent mixing in response to the reduced downward solar radiation, the CO_2 and CO concentrations decreased with height along the eastern slope. When the ambient and upslope flows converged at $\sim$ 2800 m (red z^b_{EML} in Fig. 5), once again, the upslope flow along the eastern slope transported the air with relatively high CO_2, water vapor, and aerosol concentrations from the bottom of the CBL over the plains to above the convergence altitude. The relatively stable layer, as visible between the red z^b_{EML} and $\sim$ 2600 m (marked with red dashed line in Fig. 5) in the reduced high-frequency oscillation of all the variables, was probably associated with the descending flow as part of the mountain circulation.

Because the mountain circulation was limited to the area close to the Front Range as seen from the absence of the easterly flow, the strong turbulence above 4000 m along the spiral run reflected the upwind atmospheric characteristics northwest of the run, where the air could be influenced by the mountain circulation close to the Front Range. Along the spiral run, both water vapor and CO_2 concentrations were well mixed vertically between the plains CBL top and 4000 m (Figs. 6g and h). Within this layer, the water vapor specific humidity increased from its morning value of 3 g kg $^{-1}$ to 4.5 g kg $^{-1}$. The CO_2 concentration at 2500 m increased from its morning value of 371 ppm to 374 ppm. The largest CO_2 concentration from the entire RF13 flight path was over the high mountain peaks (Fig. 7). This result indicates that ecosystem respirated and anthropogenic CO_2 were transported above the mountain peaks, and the photosynthetic reduction of the CO_2 had not reduced the relatively high CO_2 concentration at those altitudes by the time of the flight, even with the help of the daytime convective turbulent mixing and the thermally-induced mountain circulation.

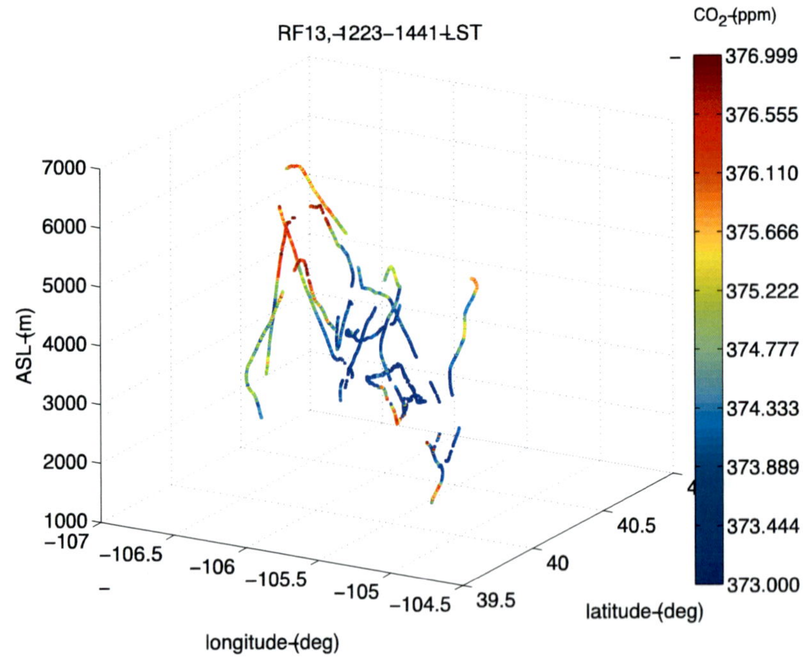

Figure 7. The CO_2 concentrations along the flight path during RF13 .

4. Conclusion

Using unique trace-gas and traditional meteorological observations on board the NCAR C-130 aircraft, we investigated interactions between thermally-induced mountain circulations and convective turbulent mixing and their impact on the transport of trace gases, such as CO_2, over the Colorado Front Range. Using the consistent observations from both traditional meteorology and known characteristics of the air associated with ecosystem activities at the bottom of the atmosphere over the plains, we identified a vertical mountain circulation over the Colorado Front Range. Because of the north-south oriented mountain ridge of the Front Range, the daytime mountain circulation east of the Front Range was found to be close to a two-dimensional vertical circulation. Bounded by mass conservation, it consisted of the upslope flow along the sunny slope west of the plains, the horizontal return flow away from the mountain and above the altitude where the upslope flow and the downslope ambient flow converge, and the descending flow over the plains. As a result of 1) the convergence between the upslope flow and the opposite ambient flow, 2) the convective mixing over the sunny slope and mountain ridge, and 3) the strong shear generated turbulent mixing close to the mountain ridge top, the air transported up by the upslope flow transitioned to a relatively deep turbulent layer above the convective boundary layer over the plains.

We found that as a result of the daytime mountain circulation, the air at the bottom of the convective boundary layer over the plains could be transported by the upslope flow to the elevated mixed layer above the convective boundary layer over the plains. We observed evidence of the upslope transport of the nighttime-accumulated moist, CO_2 rich, and polluted high-CO air over the plains. We also observed evidence of the upslope transport of the photosynthesized low-CO_2 air followed by a period of the upslope transport of high-CO_2 air resulting from reduced photosynthesis, anthropogenic pollution, and reduced turbulent mixing associated with the cloud development. Overall, the net vertical CO_2 transport by the mountain circulation throughout a day depends on the strength of the thermal forcing, the intensity of photosynthetic activity, and the magnitude of the ambient flow. Both thermal forcing and the photosynthetic activity can be strongly influenced by the diurnal variation of solar radiation and presence of clouds. Some of the ecosystem respired CO_2 can remain above the mountain ridge and remain in the lower troposphere until a strong downdraft or a strong vertical mixing event brings it down to the ground somewhere downstream.

The observations in this study imply that the net regional CO_2 ecosystem-atmosphere exchange occurs in a deep vertical and large horizontal domain and requires careful consideration of transporting and mixing processes over complex terrain. Atmospheric circulations and three-dimensional turbulent transport of CO_2 are not limited to complex terrain. Surface heterogeneity in changes of land use and land cover, for example, can lead to spatial variations of surface heating, which could also lead to significant air circulations. Monitoring CO_2 transport by both air circulations and turbulent mixing is necessary, not only over complex terrain, but also over flat terrain.

Acknowledgment

JS would like to acknowledge the support of NCAR BGS and BEACHON programs. We would like to thank Robert Banta and Beth Holland for their helpful suggestions. NCAR is sponsored by the National Science Foundation. The University Corporation for Atmospheric Research manages the National Center for Atmospheric Research under sponsorship by the National Science Foundation. Any opinions, findings, and conclusions, or recommendations expressed in this publication are those of the authors and do not necessarily reflect the views of the National Science Foundation.

References

Ahmadov, R., Gerbig, C., Kretschmer, R., Koemer, S., Neininger, B., Dolman, A. J., & Sarrat, C. (2007). Mesoscale covariance of transport and CO_2 fluxes: Evidence from observations and simulation using the WRF-VPRM coupled atmosphere-biosphere model. *J. Geophys. Res.*, **112**, D22107, doi:10.1029/2007JD008552.

Arritt, R. W., Wilczak, J. M., &Young, G. S. (1992). Observations and numerical modeling of an elevated mixed layer. *Mon. Wea. Rev.*, **120**, 2869-2880.

Aubinet, M. (2008). Eddy covariance CO_2 flux measurements in nocturnal conditions: An analysis of the problem. *Ecol. Appl.* , **18(6)**, 1368-1378.

Banta, R.M. (1984). Daytime boundary-layer evolution over mountainous terrain. Part I: Observations of the dry circulations. *Mon. Wea. Rev.*, **112**, 340-356.

Banta, R. M. (1985). Late-morning jump in TKE in the mixed layer over a mountain basin. *J. Atmos. Soc.*, **42**, 407-411.

Banta, R. M. (1986). Daytime boundary layer evolution over mountainous terrain. Part II: Numerical studies of upslope flow duration. *Mon. Wea. Rev.*, **114**, 1112-1130.

Barr, S., & Orgill, M. M. (1989). Influence of external meteorology on nocturnal valley drainage winds. *J. Appl. Meteorl. Clim.*, **28**, 497-517.

Blumen, W. (Editor), 1990: *Atmospheric Processes Over Complex Terrain.* Meteorological Monographs, 23, No. 45. American Meteorological Society, Boston, Massachusetts, USA.

Burns, S. P., & colleagues (2011). Atmospheric stability effects on wind fields and scalar mixing within and just above a subalpine forest in sloping terrain. *Boundary-Layer Meteorol.*, **138**, 231-262. doi 10.1007/s10546-010-9560-6.

Conrad, R., & Seiler, W. (1985). Influence of temperature, moisture, and organic carbon on the flux of H_2 and CO between soil and atmosphere: Field studies in subtropical regions. *J. Geophys. Res.*, **90**, 5699-5709.

De Wekker, S. F. J., Steyn, D. G., & Nyeki, S. (2004). A comparison of aerosol-layer and convective boundary layer structure over a mountain range during STAAARTE '97. *Boundary-Layer Meteorol.*, **113**, 249-271.

De Wekker, S. F. J. (2008). Observational and numerical evidence of depressed convective boundary layer heights near a mountain case. *J. Appl. Meteorl. Clim.*, **47**, 1017-1026.

Dolman, A. J., & colleagues (2006). The CARBOEUROPE regional experiment strategy. *Bull. Amer. Meteor. Soc.*, **87**, 1367-1379.

Geiger, R., Aron, R. H., & Todhunter, P. (1995). *The Climate Near the Ground.*. Vieweg, Textbook Meteorology. pp. 528.

Gudiksen, P. H., & Shearer, D. L. (1989). The dispersion of atmospheric tracers in nocturnal drainage flows. *J. Appl. Meteorl. Clim.*, **28**, 602-608.

Kalthoff, N., Horlacher,V., Corsmeier, U., Volz-Thomas, A., Kolahgar, B., Geib, H., Möllmann-Coers, M., & Knaps, A. (2000). Influence of valley winds on transport and dispersion of airborne pollutants in the Freiburg-Schauinsland area. *J. Geophys. Res.*, **105**, No. D1, 1585-1597.

Kutsch, W., Kolle, O., Rebmann, C., Knohl A., Ziegler W., & Schulze, E. D. (2008). Advection and resulting CO_2 exchange uncertainty in a tall forest in central Germany. *Ecol. Appl.* , **18(6)**, 1391-1405.

Mahrt, L., & Larsen, S. (1990). Relation of slope winds to the ambient flow over gentle terrain. *Boundary-Layer Meteorol.*, **53**, 93-102.

McGowan, H. A. (2004). Observation of anti-winds in a deep Alpine valley, Lake Tekapo, New Zealand. *Arctic, Antarctic, and Alpine Research*, **36(4)**, 495-501.

McKendry, I. G., Steyn, D. G., Banta, R. M., Strapp, W., Anlauf, K., & Pottier, J. (1998). Daytime photochemical pollutant transport over a tributary valley lake in Southwestern British Columbia. *J. Appl. Meteorl. Clim.*, **37**, 393-404.

Orgill, M. M. (1989). Early morning ventilation of a gaseous tracer from a mountain valley. *J. Appl. Meteorl. Clim.*, **28**, 636-651.

Orgill, M. M., Kincheloe, J. D., & Sutherland, R. A. (1992). Mesoscale influences on nocturnal valley drainage winds in Western Colorado valleys. *J. Appl. Meteorl. Clim.*, **31**, 121-141.

Pérez-Landa, G., Ciais, P., Sanz, M. J., Gioli, B., Miglietta, F., Palau, J. L., Gangoiti, G., & Millan, M. M. (2007a). Mesoscale circulations over complex terrain in the Valencia coastal region, Spain — Part 1: Simulation of diurnal circulation regimes. *Atmos. Chem. Phys.*, **7**, 1835-1849.

Pérez-Landa, G., & colleagues (2007b). Mesoscale circulations over complex terrain in the Valencia coastal region, Spain—Part 2: Modeling CO_2 transport using idealized surface fluxes. *Atmos. Chem. Phys.*, **7**, 1851-1868.

Reuten, C., Steyn, D. G., Strawbridge, K. B., & Bovis, P. (2005). Observations of the relation between upslope flows and the convective boundary layer in steep terrain. *Boundary-Layer Meteorol.*, **116**, 37-61.

Rotach, M. W., & colleagues (2004). Turbulence structure and exchange processes in an Alpine valley. *Bull. Amer. Meteor. Soc.*, **85**, 1367-1385.

Rotach, M., & Zardi D. (2007). On the boundary-layer structure over highly complex terrain: Key findings from MAP. *Quart. J. Roy. Met. Soc.*, **133**, 937-948.

Seinfeld, J. H., & Pandis, S. N. (2006). *Atmospheric Chemistry and Physics: From Air Pollution to Climate Change.* A Wiley-Interscience Publication, John Wiley & Sons, Inc. pp.1232.

Sun, J., & colleagues (2007). CO_2 transport over complex terrain. *Agric. Forest. Met.*, **145**, 1-21.

Sun J., & colleagues (2010). A multi-scale and multi-disciplinary investigation of ecosystem-atmosphere CO_2 exchange over the Rocky Mountains of Colorado. *Bull. Amer. Meteor. Soc.*, **91(2)**, 209-230.

Tarr, A. M., Miller, W. L., & Zepp, R. G. (1995). Direct carbon monoxide photoproduction from plant matter. *J. Geophys. Res.*, **100**, 11,403-11,413.

Wallace, J. M., & Hobbs, P. V. (2006). *Atmospheric Science: An Introductory Survey*. Elsevier Academic Press. pp. 480.

Whiteman, C. D. (1982a). Breakup of temperature inversions in deep mountain valleys: Part I. Observations. *J. Appl. Meteorl. Clim.*, **21**, 270-289.

Whiteman, C. D. (1982b). Breakup of temperature inversions in deep mountain valleys: Part II. Thermodynamic model. *J. Appl. Meteorl. Clim.*, **21**, 290-302.

Whiteman, C. D. (1989). Morning transition tracer experiments in a deep narrow valley. *J. Appl. Meteorl. Clim.*, **28**, 626-635.

Whiteman, C. D. (2000). *Mountain Meteorology—Fundamentals and Applications*. Oxford University Press, p. 355.

Wolff, J. K., & Sherman, R. D. (2008). Climatology of upper-level turbulence over the contiguous United States. *J. Appl. Meteorl.*, **47**, 2198-2214.

Reviewed by Robert M. Banta from Earth System Research Laboratory, National Oceanic and Atmospheric Administration, Boulder, Colorado, USA; and Elisabeth Holland from Atmospheric Chemistry Division, Earth System Laboratory, National Center for Atmospheric Research, Boulder, Colorado, USA.

INDEX

C

D

E

F

G

H

I

K

L

M

N

O

P

T

U

V

W

Z